Glycerol

Claudio J.A. Mota • Bianca Peres Pinto
Ana Lúcia de Lima

Glycerol

A Versatile Renewable Feedstock for the Chemical Industry

Claudio J.A. Mota
Institute of Chemistry, School of Chemistry
INCT of Energy and Environment
Federal University of Rio de Janeiro
Rio de Janeiro, Rio de Janeiro, Brazil

Bianca Peres Pinto
Institute of Chemistry
Federal University of Rio de Janeiro
Rio de Janeiro, Rio de Janeiro, Brazil

Ana Lúcia de Lima
Institute of Chemistry
Federal University of Rio de Janeiro
Rio de Janeiro, Rio de Janeiro, Brazil

ISBN 978-3-319-86611-6 ISBN 978-3-319-59375-3 (eBook)
DOI 10.1007/978-3-319-59375-3

Softcover reprint of the hardcover 1st edition 2017

Printed on acid-free paper

This Springer imprint is published by Springer Nature
The registered company is Springer International Publishing AG
The registered company address is: Gewerbestrasse 11, 6330 Cham, Switzerland

Preface

The industrial revolution was based on coal, which flourished as the major source of energy and raw material for the chemical industry in the nineteenth century. With the beginning of the commercial exploration of oil fields in the USA, in 1859, the world's energy matrix was gradually changed, and the twentieth century may be defined as the petroleum era. The use of oil as the major source of energy also brought the petrochemical industry, with the production of plastics, which gradually replaced wood, metal and other natural resources in human civilization. At the beginning of the twenty-first century, the world experiences a transition. Oil, coal and natural gas will still play a major role in the energy matrix and production of chemicals, but the scenario must gradually move towards more sustainable energy sources. Climate changes are one of the biggest issues of these days and may bring great economical, environmental and social impacts. The crescent emissions of greenhouse gases since the beginning of the industrial revolution have turned on the yellow light, making the whole humanity think about the destiny of the Earth, up to now, our only home in the vast universe.

Biofuels can be a response to global warming and may help in decreasing carbon emissions. Nevertheless, there are still many developments to be taken, before biofuels could replace a significant part of the oil and coal consumed worldwide. In addition, it will not be the only source of renewable energy in the future, but together with solar, wind, tide and nuclear (based on fusion processes) may provide a more diversified and sustainable energy matrix up to the end of this century.

Bioethanol and biodiesel are the main biofuels used worldwide at the beginning of the twenty-first century. Bioethanol is based on carbohydrate processing, whereas biodiesel relies on the transformation of triglycerides. Both biofuels are facing increased production and consumption in recent years, and the forecast points out for a continuous grow. Biodiesel is produced through the transesterification of vegetable oils and animal fat. Used cooking oils, as well as sewer residues, may also be used. Algae appear as a potential biomass source of biodiesel in the future. Therefore, it is expected that many different raw materials may account for the biodiesel production in the forthcoming years. Nevertheless, no matter the source of raw material, glycerol will always be formed as a by-product of transesterification of triglycerides.

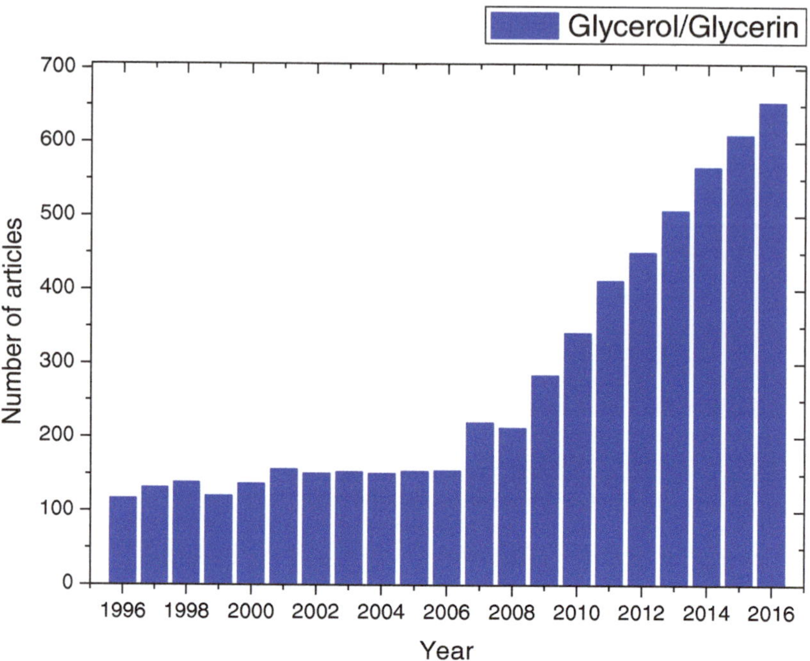

Fig. 1 Number of published scientific articles with glycerol or glycerin as keywords

Hence, its proper valorization will be extremely important to the economic, environmental and social feasibility of the biodiesel, making the whole production chain more sustainable.

This book aims to cover the progresses made in the glycerol conversion to commodities and specialty chemicals. Since the beginning of the widespread use of biodiesel, in the 1990s, there appears a great number of works on the use of glycerol, especially related with its chemical transformation. Figure 1 shows the number of scientific papers published, since 1996, having glycerol or glycerin as keyword. Great interest in the topic is clear, especially in the past 10 years, when the number of papers has skyrocketed. On the other hand, although there have been some review articles published within this time, there is only one book on the use of glycerol that was published about 10 years ago. Thus, taking into account the importance and actuality of the subject, an updated book, covering the major advances in the chemical transformation of glycerol, is highly interesting. This is the major objective of this contribution.

In Chap. 1, the biofuel scenario is briefly discussed, with emphasis on the biodiesel production chain. The chapter also includes discussion on glycerol as a by-product of the biodiesel production, together with other sources of glycerol, such as the chemical synthesis from propene and from algae.

Chapter 2 highlights the utilization of glycerol with a brief discussion on traditional uses, such as in personal care products, in pharmaceuticals and in the food industry. A historic perspective is also included, covering the first identification by the Swedish chemist Carl Scheele and the widespread utilization of nitroglycerine after the development of dynamite by Alfred Nobel. Other uses, such as animal food supplement, in energy generation through combustion or transformation in biogas and in formulations of fluids for enhanced oil recovery are briefly addressed too. The chapter points out the great potential of glycerol as a renewable raw material to the chemical industry, highlighting some potential transformations, most of them discussed in more details in the following chapters. The problem of glycerol purity is also addressed in Chap. 2.

The biotechnological processes of glycerol transformation in high-value chemical products are discussed in Chap. 3. An introduction to the growing importance of the biotechnological processes in the chemical industry is briefly addressed, highlighting the focus on specialty chemicals instead of commodities. Several processes of biochemical transformation of glycerol are then discussed, covering the production of 1,3-propanediol, an important chemical used in the production of textile fibres, ethanol, citric, lactic, succinic and propionic acids, as well as hydrogen and dihydroxyacetone. This later product is widely used in formulations of self-tanning lotions, having a high added-value compared with glycerol.

Chapter 4 is dedicated to the production of commodities from glycerol through thermochemical processes. The chapter begins with a brief discussion on the structure of the chemical industry, showing the three generations of the so-called petrochemical sector. A discussion on the production of renewable ethylene, from ethanol dehydration, and the ethylene-based tree of products is used to address the potential of glycerol as a renewable raw material to the propylene-based tree of products. The chapter covers studies on glycerol hydrogenolysis to propylene glycol, which is presently a commercial process, as well as to propene, which is not yet fully studied, but has great industrial appeal. Glycerol chlorination to epichlorohydrin is another commercial process that shows the potential of this by-product of biodiesel production in the chemical industry. Some aspects of this process in relation to the traditional one, based on fossil sources, are briefly addressed. Glycerol dehydration is also discussed, with emphasis on the production of acrolein and acrylic acid. The chapter ends reviewing studies on glycerol reforming to produce syngas and hydrogen, as well as discussing some developments to transform glycerol in methanol.

The production of specialty chemicals from glycerol via thermochemical conversions is highlighted in Chap. 5. Glycerol derivatives, such as ethers, acetals/ketals and esters, have been extensively studied in the past 10 years, showing great potential as fuel additives. The processes to produce these derivatives and their utilization in the fuel sector are consistently discussed. The chapter also covers the production of glycerol carbonate, a relatively new product that is finding crescent applications every year. The thermodynamic limitations of direct glycerol carbonation with CO_2 are briefly discussed, together with other processes based on the reaction of glycerol with urea. Finally, the oxidation of glycerol to dihydroxyacetone, glyceric and mesoxalic acids, among other products, is addressed in Chap. 5.

The last chapter discusses the conversion of glycerol to specialty and commodity chemicals in the biorefinery context. The chapter begins with a brief comparison between a traditional oil refinery and a biorefinery, based on the conversion of renewable biomass raw materials. The production of specialty chemicals from glycerol is addressed with emphasis on glycerol ethers and acetals/ketals, with high potential to be used as fuel additives, glycerol carbonate and dihydroxyacetone. The main discussions are concerned with the whole sustainability and feasibility of the processes at industrial scale. The chapter ends with a discussion on glycerol hydrogenolysis to propene and glycerol dehydration to acrolein and acrylic acids, as examples of biorefinery processes to commodities. A contextualization in terms of logistics, potential profitability and drawbacks relative to the traditional processes, based on fossil sources, is addressed.

We believe that the book may be a good and updated source of consult to researchers, students and professionals from academy, industry and the government. Our aim is to provide concise information on this subject, but also to cover broad aspects and to include relevant references for further consult of the interested reader.

Rio de Janeiro, Brazil
February 28, 2017

Claudio J.A. Mota
Bianca Peres Pinto
Ana Lúcia de Lima

Contents

Chapter 1
Biomass and Biofuels

Abstract Biofuels are the new frontier for vehicular motion. They may be used in blends with traditional fuels, like gasoline and diesel, or neat in some circumstances. Bioethanol and biodiesel are, presently, the most commonly used biofuels in the world. They are primarily produced from edible biomass sources, but new generation of biofuel technologies will provide access to these biofuels from nonedible biomass sources. The chapter discusses the main generations of biofuel production, with particular emphasis in the technologies and raw materials of biodiesel production, as well as the glycerol produced as by-product. A general view of the world market and production of biodiesel, together with the characteristics of the glycerol obtained, is presented, as well as other sources of glycerol production, such as from sugar fermentation, algae and fossil raw materials.

Keywords Biofuels • Energy • Biodiesel • Glycerol • Bioethanol • Pyrolysis

1.1 Biofuels

Since the beginning of the industrial revolution, the concentration of carbon dioxide in the atmosphere has increased by 40%, surpassing 400 ppm in 2014. Carbon dioxide is a greenhouse gas, being associated with global warming and climate changes. During the last United Nation meeting for Climate Change (COP-21), in 2015, in Paris, nations have agreed to keep the temperature increase by 1.5°C above the pre-industrial era by the end of this century. This will require a dramatic reduction of their carbon footprint and use of fossil fuels, to reduce the emission of CO_2 and slow down global warming.

In this scenario, the use of biofuels will increase in importance in the forthcoming years. First-generation biofuels rely on food-derived biomass (Fig. 1.1), the most important being bioethanol, usually produced from sugar cane, corn or sugar beet and biodiesel, which is usually fabricated from vegetable oils and fats (soybean, rapeseed, palm or tallow) [1]. Biobutanol, produced from sugar fermentation, can also be considered as first-generation biofuel, but its commercial importance is still significantly lower than bioethanol and biodiesel. Second-generation biofuels are growing fast, being based in nonedible biomass sources. Thus, bioethanol can be produced from cellulosic materials [2], upon the controlled hydrolysis of

C.J.A. Mota et al., *Glycerol*, DOI 10.1007/978-3-319-59375-3_1

polysaccharides (cellulose and hemicellulose), whereas biodiesel could be produced from nonedible oils, such as *Jatropha curcas*. Third-generation biofuels are also based on lignocellulosic material and involve processes such as pyrolysis and gasification [3], among others, to produce a great range of fuels. Some people also consider the fourth-generation biofuels, which are primarily based on algae, which requires significantly lower areas for cultivation, but still faces many challenges to reduce production costs.

Figure 1.2 shows a simplified scheme of biomass transformation. The direct burning of biomass, like sugar cane bagasse, can yield heat to produce electricity. Hydrolysis and fermentation of sugars yield bioethanol, whereas extraction of triglycerides allows the production of biodiesel, either by transesterification or hydrogenolysis (renewable diesel). Lignocellulose materials can be directly converted into bio-oil through pyrolysis, whereas gasification followed by Fischer–Tropsch synthesis can yield gasoline and diesel-range hydrocarbons, as well as methanol with the use of Cu-ZnO catalysts.

The USA and Brazil are the world's leaders in the production of biofuels, followed by Germany and China (Fig. 1.3). Bioethanol has a production capacity far larger than biodiesel, as well as different major sources. In Brazil, sugar cane is the main crop used in the production of bioethanol, whereas in the USA, corn is the major source. Bioethanol is, today, the major biofuel used worldwide, replacing methyl *tert*-butyl ether (MTBE) as the main oxygenated additive in gasoline. The

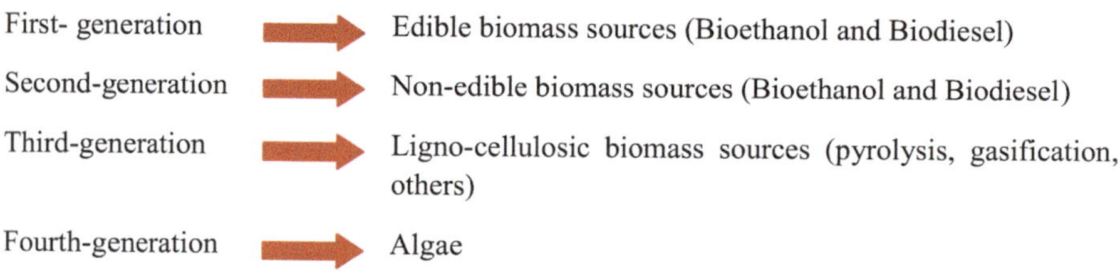

Fig. 1.1 The four generations of biofuels

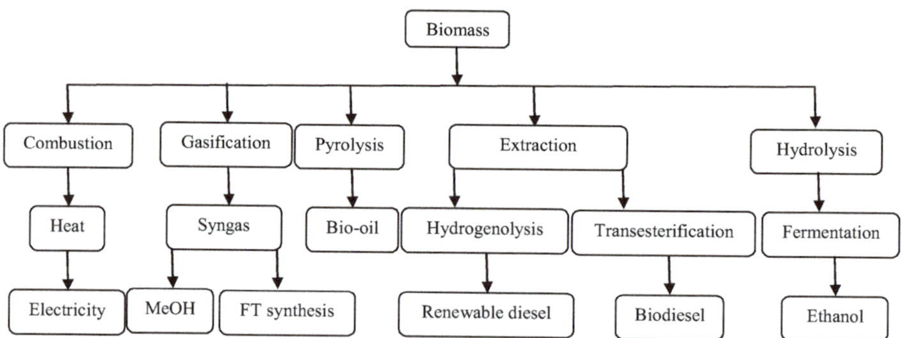

Fig. 1.2 Simplified scheme of biomass upgrading into energy and fuels

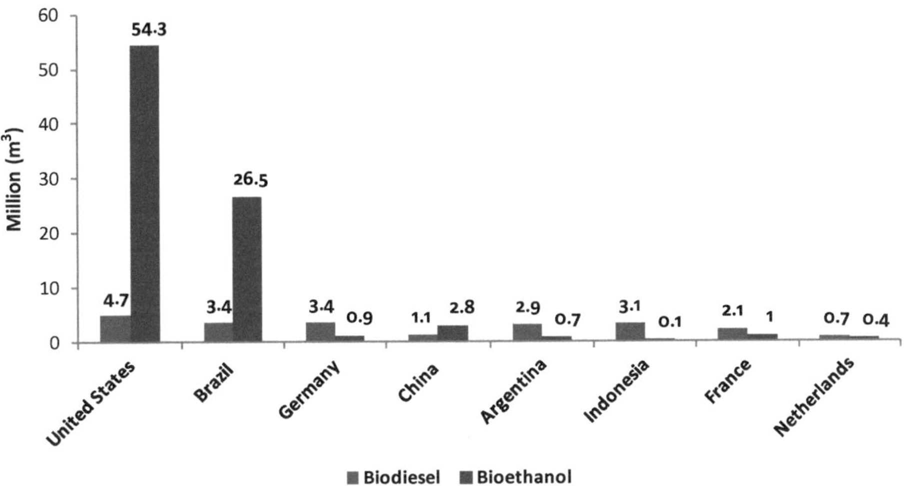

Fig. 1.3 Production of biofuels in the world in 2015 (Source: Brazilian Energy Research Enterprise, 2015)

feedstock to produce biodiesel also varies with the region; soybean oil is the major raw material in Brazil and the USA, whereas rapeseed oil is more important in Europe and palm oil in Asia.

1.2 Biodiesel

Biodiesel is usually composed of methyl esters of fatty acids, produced upon transesterification of triglycerides under basic, acid or enzymatic catalysis [4]. Other processes of obtaining diesel-range fuels include cracking and hydrogenolysis of triglycerides [5]. However, these processes produce mainly hydrocarbons, which are usually referred as renewable diesel.

Transesterification is an equilibrium reaction where the triglyceride reacts with a short-chain alcohol, usually methanol, in the presence of a basic catalyst affording three molecules of fatty acid methyl esters (FAME) and one molecule of glycerol (Scheme 1.1). Equilibrium is shifted towards products upon using an excess of alcohol. In addition, glycerol is not soluble in the oil or in the biodiesel. Thus, phase separation occurs favouring the progress of the reaction. Although acid and enzymatic catalysis may be used in the transesterification of oils and fats, base catalysis is predominant in a commercial scale. The main reason is the mild reaction conditions, with temperatures around 70°C and 60 min of reaction time. Today, sodium and potassium hydroxides, as well as sodium methoxide, are the main basic catalysts used to produce biodiesel. Nevertheless, they are difficult to be recovered at the end of the process and remain mostly dissolved in the glycerol phase. The development of heterogeneous basic catalysts is a topic of great interest [6], because of

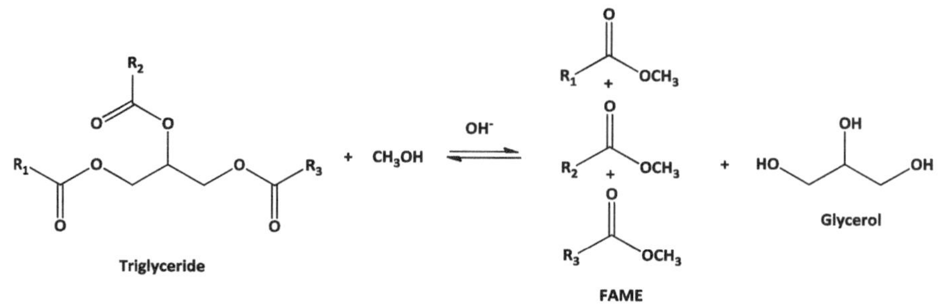

Scheme 1.1 Transesterification of triglycerides with methanol to afford biodiesel (FAME)

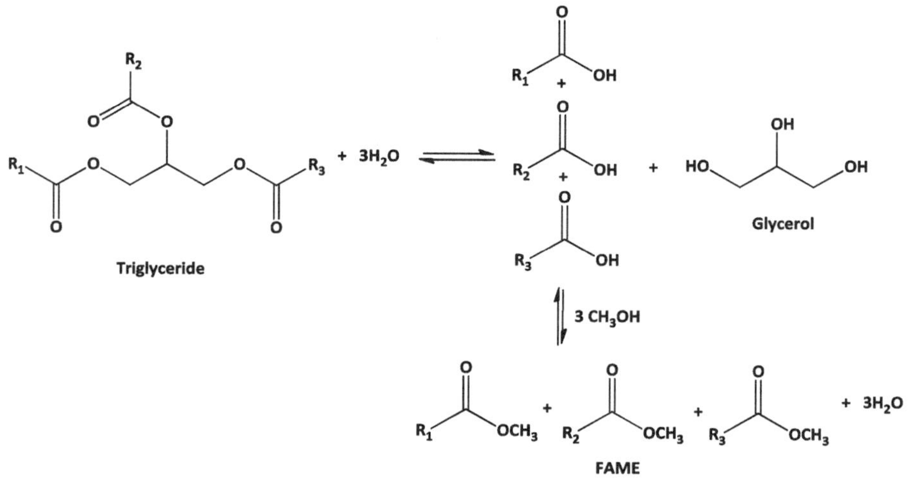

Scheme 1.2 Hydroesterification process of biodiesel production

the potential reutilization, use in continuous flow processes and production of a purer glycerol phase.

Fatty acids are by-products of vegetable oil refining and may also be a source of biomass to produce biodiesel. In this case, the main process involves the acid-catalysed esterification with methanol and does not produce glycerol as by-product. In fact, the presence of free fatty acids in the vegetable oil limits the use of basic catalysts. In general, acidity of 1% or higher requires the pretreatment of the oil or the use of alternative processes. Hydroesterification may be an option and involves the initial hydrolysis of the triglycerides, followed by esterification of the fatty acid formed (Scheme 1.2). The first reaction can be catalysed by acids or enzymes and yields purer glycerol phases.

In 2015, the world's production of biodiesel accounted for nearly 24 billion litres. Figure 1.4 shows the main producers, with the USA, Brazil and Germany in the top three positions. In Europe, rapeseed is the main source of biomass to produce biodiesel, whereas in the USA, Brazil and Argentina, soybean is predominant.

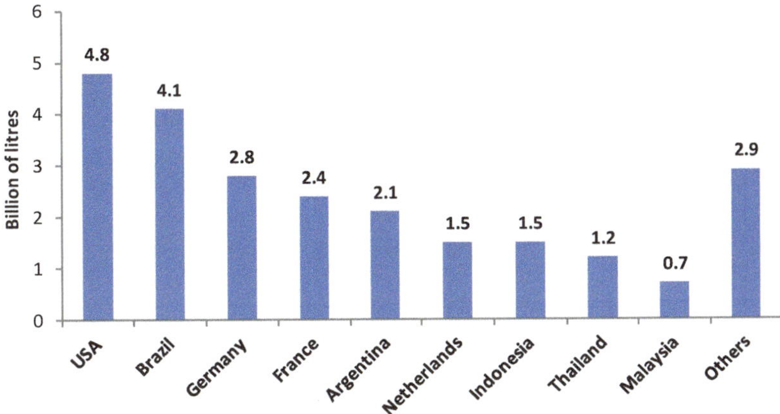

Fig. 1.4 World's biodiesel production by country in 2015 (Source: Statista—own elaboration)

Fig. 1.5 Main raw materials for the production of biodiesel: (**a**) soybean, (**b**) palm, (**c**) rapeseed, (**d**) coconut, (**e**) sunflower, (**f**) cotton, (**g**) peanut, (**h**) corn, (**i**) *Jatropha curcas*, (**j**) used cooking oil, (**k**) algae, (**l**) tallow. (Reproduced from Ref. [6] with permission of the Royal Society of Chemistry)

In Asia, palm has a major role in the production of biodiesel. Other oils, including used cooking oil, have secondary importance. Around 15% of the Brazilian biodiesel comes from tallow. Figure 1.5 shows some potential raw material used in the production of biodiesel.

The world's market of vegetable oil is presently around 180 million ton per year. Production has been increasing at an annual rate of 5% since 2013. Four major oils account for 85% of the total production. Palm is the world's most produced oil, sharing 35% of the market, being used as the main vegetable oil in the food industry. Soybean and rapeseed respond for 26% and 15% of the oil market, respectively, whereas sunflower accounts for 9% of the total sales.

Table 1.1 Average composition of some vegetable oils and fats

	Acids				
Oil	Palmitic 16:0[a]	Stearic 18:0[a]	Oleic 18:1[a]	Linoleic 18:2[a]	Others
Palm	47	6.5	36	6.5	4
Rapeseed	3.5	1	54	22	19.5
Soybean	11	2	20	64	3
Corn	6	2	44	48	
Peanut	8.5	6	51.5	26	
Tallow	29	24.5	44.5		

[a]Numbers refer to the carbon atoms and the double bonds in the hydrocarbon chain, respectively

Biodiesel properties depend on the nature of the hydrocarbon chain, which in turn is associated with the raw material used. Table 1.1 shows the average distribution of fatty acids in several oils and in tallow. Palm and tallow are rich in saturated fatty acids, whereas soybean oil has predominantly polyunsaturated chains. The oxidative stability of the biodiesel is associated with the degree of unsaturation; as the number of double bonds in the chain increases, the biodiesel is more prone to be oxidized, requiring the addition of antioxidant additives. On the other hand, saturated fatty acid chains decrease the cold flow properties of the biodiesel, increasing the pour, freezing and flow temperatures. Therefore, additives are required if the biodiesel will be used in low-temperature conditions.

The production of biodiesel generates economic, environmental and social benefits [7]:

- It is renewable, biodegradable, non-toxic, non-flammable and free of sulphur and aromatics.
- Emits less hazardous gases, such as CO and SO_2, particulate matters and hydrocarbons compared to diesel.
- Helps to comply with the global agreements to reduce greenhouse gas emissions.
- Features high flash point (above 100°C) making it safer to be transported, handled, distributed and used.
- Does not need additional lubricants, like diesel fuels.
- Presents higher cetane number compared with average diesel fuels.
- Does not require engine modification at low rates of blending. Minor modifications are required for blending over 20%.
- Promotes rural development, reduces dependence on diesel and helps keeping human resources in the field.

1.3 Glycerol of Biodiesel Production

From the stoichiometry of the transesterification of triglycerides, glycerol is formed in about 10 wt% mass balance. The crude glycerol of biodiesel production has water, methanol and dissolved salts as major impurities; other components, such as mono- and diacylglycerides, can also be present in small amounts. Table 1.2 shows the average composition of the crude glycerol, also named glycerine, from a Brazilian biodiesel plant. The sodium chloride is formed upon neutralization of the glycerol phase with HCl, due to the presence of the dissolved basic catalyst.

The glycerol of biodiesel production can be refined for further applications. After transesterification, soap may be formed and remains dissolved in the glycerol phase. Hence, acid treatment is necessary to transform soap into free fatty acids, which can be easily separated from the top because they are not soluble in glycerol. Excess methanol can be recovered by distillation to be reused in the transesterification process. The salts remaining in the glycerol phase are deleterious impurities that limit the use of glycerol in some chemical processes [8]. Purification is required to achieve a product with the necessary specification for further uses.

1.4 Other Sources of Glycerol

Although in recent years the biodiesel production has been the major source of glycerol, this chemical can also be produced from other routes and raw materials.

Traditionally, glycerol is a by-product of the soap industry. Hydrolysis of fats and vegetable oils affords carboxylate salts (soap) and glycerol. Figure 1.6 shows a simplified process diagram of soap production.

Table 1.2 Average composition of crude glycerol from a Brazilian biodiesel plant

Composition	Wt%
Glycerol	80
Water	7
Methanol	1
NaCl	12

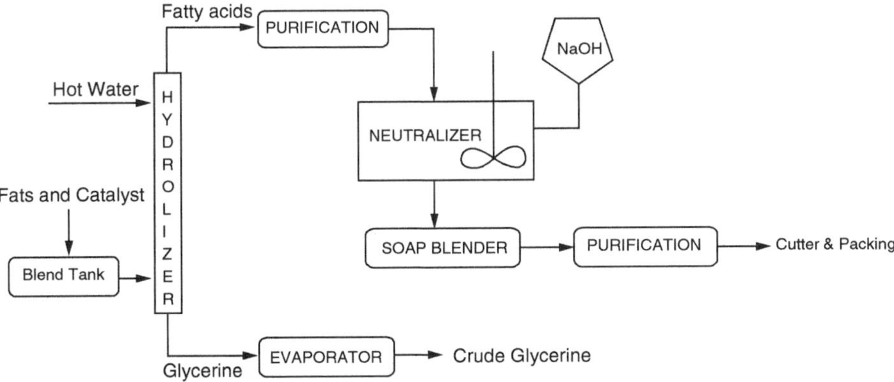

Fig. 1.6 Simplified diagram of soap-producing process

A process to convert propene into glycerol was developed at the end of the First World War. The driven force was the crescent use of oil at that time and the birth of the petrochemical industry. Today, this process is being phased out due to the surplus of glycerol from biodiesel production. Scheme 1.3 shows the reaction sequence from propene to glycerol. The first step is chlorination at high temperatures, involving free radicals, to yield allyl chloride and HCl. Then, addition of hypochlorous acid produces a mixture of 1,3-dichloro-2-propanol and 2,3-dichloro-1-propanol, which upon treatment with excess NaOH yields glycerol and chlorinated residues.

The process is conceived to have three different reaction sections. In the first reactor, there occurs the chlorination of propene to yield allyl chloride and HCl. After quenching and separating unreacted propene and gaseous HCl, the effluent is treated with hypochlorous acid to yield the halohydrin isomers. The final reactor involves the hydrolysis with 10% NaOH solution, followed by crystallization of the NaCl formed and purification of the glycerol (Fig. 1.7).

Glycerol can be obtained from algae [9]. Some species of the unicellular algae *Dunaliella* possess outstanding adaptability and tolerance towards a wide range of salinities from seawater. The capability of the cell to thrive in high salt concentrations depends on its unique ability to produce intracellular glycerol [10], which counterbalances the high external osmotic stress. *Dunaliella salina* and *Dunaliella viridis* grow in media containing different salt concentrations (Fig. 1.8). Algae growth and glycerol production increase as the salinity of the medium increases [11]. Typically, *Dunaliella* can grow in media with NaCl concentration of 4 mol L^{-1} and higher, as the Dead Sea, in the Middle East. Besides the sunlight, the alga needs CO_2 and nutrients, such as nitrates, phosphates and traces of metals for growing, at temperatures ranging from 10 to 40°C and pH between 7 and 9. At optimized conditions, the intracellular medium contains between 25 and 30% of glycerol.

Glycerol may be also formed during sugar fermentation. In the ethanol-producing process from glucose, glycerol is formed in about 3.5 wt%, from the reduction of the $NADH^+$ with the sugar molecules. Addition of Na_2SO_3 increases the glycerol production, because sulphite complexes with acetaldehyde inhibit the ethanol formation [12]. Fermentation at high pH also increases the glycerol yield [13].

Propene $\xrightarrow[500\,^\circ C]{Cl_2}$ HCl + Cl–CH₂–CH=CH₂ $\xrightarrow{HOCl}$ Cl–CH₂–CH(Cl)–CH₂–OH + Cl–CH₂–CH(OH)–CH₂–Cl $\xrightarrow{2\,NaOH}$ HO–CH₂–CH(OH)–CH₂–OH (Glycerol) + 2 NaCl

Scheme 1.3 Industrial route of glycerol production from propene

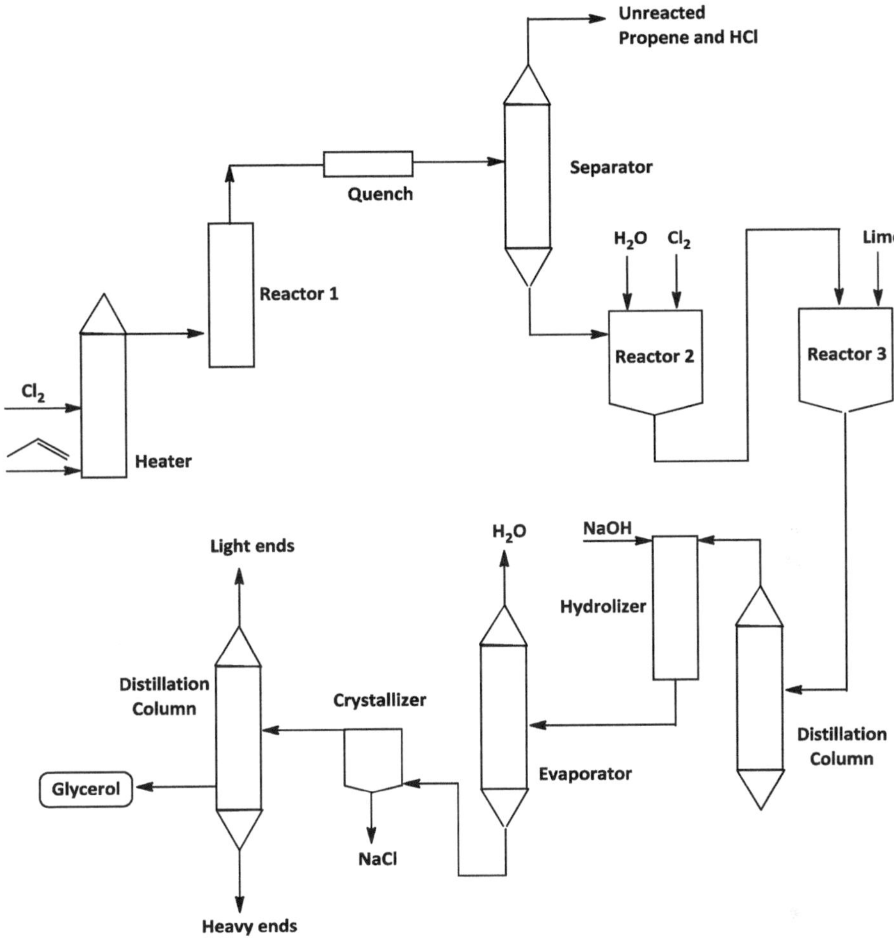

Fig. 1.7 Simplified flow diagram of the industrial process of glycerol production from propene

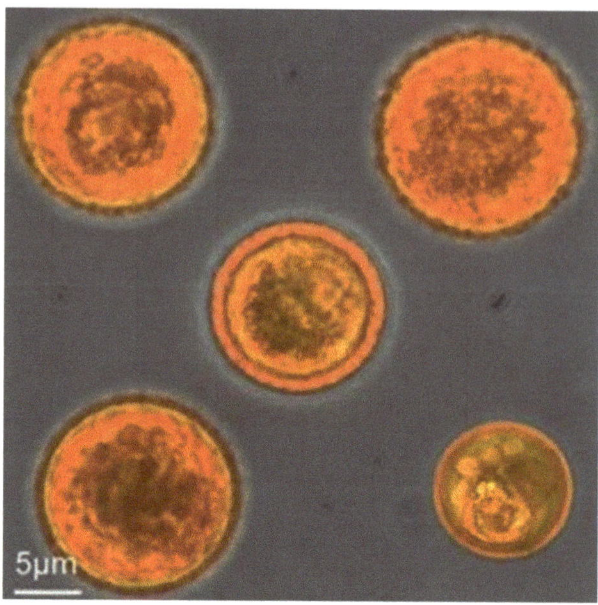

Fig. 1.8 *Dunaliella salina* at the microscope (Source: https://br.pinterest.com/mikkoutriainen/dunaliella-salina/)

References

1. Saladini F, Patrizi N, Pulselli FM, Marchettini N, Bastianoni S (2016) Guidelines for energy evaluation of first, second and third generation biofuels. Renew Sustain Energy Rev 66:221–227
2. Aditiya HB, Mahlia TMI, Chong WT, Hadi Nur AHS (2016) Second generation bioethanol production: a critical review. Renew Sustain Energy Rev 66:631–653
3. Damartzis T, Zabaniotou A (2011) Thermochemical conversion of biomass to second generation biofuels through integrated process design: a review. Renew Sustain Energy Rev 15:366–378
4. Ma F, Hanna MA (1999) Biodiesel production: a review. Bioresour Technol 70:1–15
5. Helwani Z, Othman MR, Aziz N, Fernando WJN, Kim J (2009) Technologies for production of biodiesel focusing on green catalytic techniques: a review. Fuel Process Technol 90:1502–1514
6. Lima AL, Ronconi CM, Mota CJA (2016) Heterogeneous basic catalysts for biodiesel production. Cat Sci Technol 6:2877–2891
7. Anuar MR, Abdullah AZ (2016) Challenges in biodiesel industry with regards to feedstock, environmental, social and sustainability issues: a critical review. Renew Sustain Energy Rev 58:208–223
8. Silva CXA, Mota CJA (2011) The influence of impurities on the acid-catalyzed reaction of glycerol with acetone. Biomass Bioenergy 35:3547–3551
9. Muscatine L (1967) Glycerol excretion by symbiotic algae from corals and Tridacna and its control by the host. Science 156:516–519
10. Chitlaru E, Pick U (1991) Regulation of glycerol synthesis in response to osmotic changes in dunaliella. Plant Physiol 96:50–60
11. Hadi MH, Shariati M, Afsharzadeh S (2008) Microalgal biotechnology: carotenoid and glycerol production by the green algae *Dunaliella* isolated from the Gave-Khooni salt marsh, Iran. Biotechnol Bioprocess Eng 13:540–544
12. Neuberg C, Reinfurth E (1918) Natiirliche und erzwungene Glycerinbildung beider alkoholischen Garung. Biochem Z 92:234–266
13. Connstein W, Ludecke K (1919) Uber Glyzeringewinnung durch Garung. Ber Deut Chem Gesel 52:1385–1391

Chapter 2
Glycerol Utilization

Abstract Glycerol is traditionally used in soaps, cosmetics, personal care products, pharmaceuticals and food products. All these sectors cannot drain the increasing surplus of glycerol of biodiesel production. Other applications, such as supplement for animal food, fermentation to biogas and formulations of fluids for enhanced oil recovery are gaining importance in the past years. Nevertheless, the most promising application is in the chemical industry, as a cheap and renewable raw material for the production of polymers and specialty chemicals. This chapter briefly describes the traditional and potential new applications of glycerol, also highlighting the different glycerol sources and grades, as well as the main purification procedures.

Keywords Glycerol • Glycerine • USP • Dynamite • Pharmaceutical • Chemical industry

2.1 Glycerol Properties

Glycerol is the 1,2,3-propanetriol. It was firstly identified in 1779 by the Swedish chemist Carl Wilhelm Scheele (Fig. 2.1), upon heating olive oil with litharge (PbO). The viscous liquid that separated from the oil phase was named glycerol due to its sweet taste (from the Greek glykos = sweet). The term glycerine or glycerin applies to commercial products, which have at least 95 wt% of glycerol. Some properties of glycerol are shown in Table 2.1. It is a polar, viscous, transparent liquid at ambient temperature, soluble in water and polar media and insoluble in hydrocarbons and other non-polar media. Although the melting point is near 18°C, small amounts of dissolved water impair the crystallization of glycerol, which remains as liquid at significantly lower temperatures. An eutectic mixture of two parts of glycerol and one part of water freezes at −46.5°C.

C.J.A. Mota et al., *Glycerol*, DOI 10.1007/978-3-319-59375-3_2

Fig. 2.1 Carl Wilhelm Scheele (Source: https://br.pinterest.com/pin/318629742356153831)

Table 2.1 Selected glycerol properties

Description value	Description value
Molecular formula	$C_3H_5(OH)_3$
Molecular weight (g)	92
Melting point (°C)	17.8
Boiling point (°C)	290
Viscosity (Pa s^{-1})	1.5
Vapour pressure at 20°C (mmHg)	<1
Density at 20°C (g mL^{-1})	1.261
Flash point (°C)	160 (closed cup)
Residue on ignition (wt%)	<0.01
Heat of fusion at 18.07°C (cal g^{-1})	47.49
Auto-ignition temperature (°C)	400
Critical temperature (°C)	492.2
Critical pressure (atm)	42.5
Surface tension (N m^{-1})	64,000
Specific heat (cal g^{-1} $°C^{-1}$)	0.579
Heat of combustion (kcal mol^{-1})	397
Dielectric constant (25°C)	44.4
Dipole moment (Debye)	2.7

2.2 Traditional Uses of Glycerol

One of the first uses of glycerol was in the form of trinitrate. In 1846, Ascanio Sobrero treated glycerol with a mixture of concentrated nitric and sulphuric acids, at low temperatures, isolating an oily liquid, the glycerol trinitrate or simply nitroglycerine (Scheme 2.1). The explosive nature of this compounds attracted great interest, but it was only Alfred Nobel's (Fig. 2.2) discovery in 1863, which blasting could be

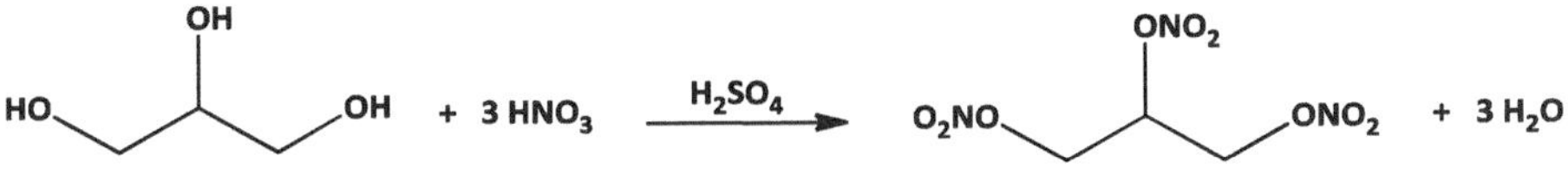

$$HOCH_2CH(OH)CH_2OH + 3\,HNO_3 \xrightarrow{H_2SO_4} O_2NOCH_2CH(ONO_2)CH_2ONO_2 + 3\,H_2O$$

Scheme 2.1 Synthesis of nitroglycerine

Fig. 2.2 Alfred Nobel (Source: https://www.nobelprize.org/alfred_nobel/biographical/articles/life-work/)

controlled upon mixing nitroglycerine with silica, that it found commercial application with the name of dynamite. Today, it is still the major explosive used in constructions and demolitions but also finds application in medicine as a vasodilator.

Glycerol is widely used in food and beverages, where it acts as humectant, sweetener and preservative. It may indirectly be used as monoacylglycerides, which are emulsifiers and stabilizers of many food products. Cheese, yogurt, powdered milk, condensed milk, precooked pasta, rolled oats, breakfast cereals, rice or tapioca pudding, breads or batters are among the food products that may contain glycerine. Some alcoholic, cider and flavoured beverages may contain glycerine, which can be also present in sauces, vinegar, mustard, condiments, tabletop sweeteners, butter and nut butter formulations as well as in candies and soups. Glycerine may provide 4.32 kcal/g of food-energy value.

Glycerine serves as solvent, moistener and humectant in drug and pharmaceutical applications, such as elixirs, ointments and capsules for medicinal use. Glycerine suppositories are recommended for treating constipation in children and adults. Glycerine is also used as emollient and demulcent in skin preparations. It may also find application in veterinary drugs.

Perhaps the most well-known application of glycerine is in cosmetics and toiletries. It acts as humectant, emollient and vehicle, being a major ingredient in toothpastes, skin creams, lotions, shaving, deodorants and make up products. Glycerine is a major component of fine soaps (Fig. 2.3).

The growth of the biodiesel industry led to a large surplus of glycerol, significantly affecting its price. Actually, the market price of crude glycerol (80% purity)

Fig. 2.3 Glycerine soap (Source: https://br.pinterest.com/pin/518617713318791605/)

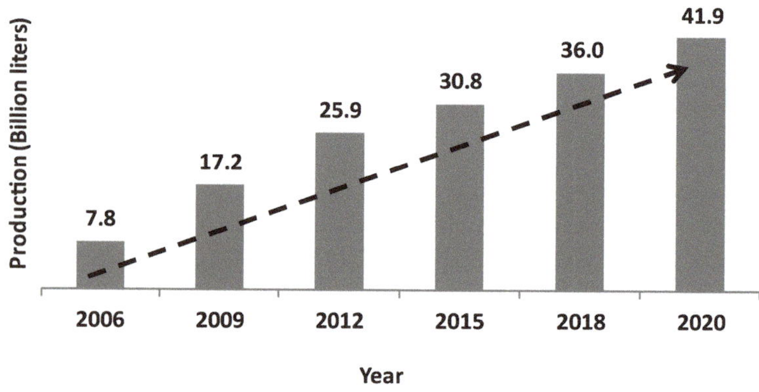

Fig. 2.4 World's scenario of crude glycerol [1]

is between US$0.09 and US$0.20 per kg, whereas the pure glycerol price ranges from US$0.60 to US$0.91 per kg. The projections for 2020 predict a world's glycerol production of about 41.9 billion litres (Fig. 2.4). Thus, the use of glycerol is extremely important to the sustainability of the biodiesel industry [1].

Figure 2.5 shows many of the present applications of glycerol. Besides the traditional uses in the pharmaceutical and personal care sectors, glycerol of biodiesel production is finding applications as supplement for animal food and fluids for enhanced oil recovery. Because of the sweet taste of glycerol, it is well accepted by animals and provides additional energy. It has been reported that glycerol may prevent ketosis on dairy cows [2]. A drawback is the presence of methanol, which, in many countries, is prohibited to be part of animal food formulations.

Crude glycerol of biodiesel production has been used in fluids for enhanced oil recovery [3]. Nevertheless, this use is limited by logistics, because the oil fields must be in close proximity to the biodiesel plants, to avoid excessive expending with transportation.

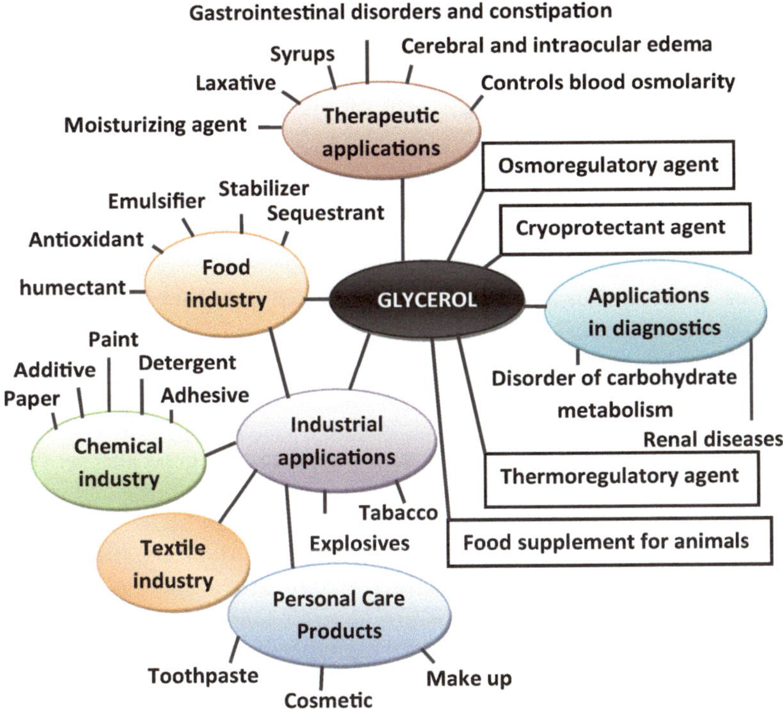

Fig. 2.5 Different applications of glycerol

$$2\ \text{HOCH}_2\text{CH(OH)CH}_2\text{OH} + 7\,O_2 \longrightarrow 6\,CO_2 + 8\,H_2O \quad \Delta H = -386\ \text{kcal/mol}$$

Scheme 2.2 Glycerol combustion

Another approach of using the glycerol of biodiesel production is in the energy sector. The combustion of glycerol (Scheme 2.2) can provide some of the heat required to run the biodiesel plant. Nevertheless, care must be taken to avoid incomplete combustion and formation of acrolein, which is highly toxic. This is specially favoured at temperatures between 200 and 300°C. In addition, the glycerine of biodiesel production may not be able to be used without some purification step, because of the high content of dissolved salts, which may corrode and damage the equipments.

An alternative approach of powering the biodiesel plant with glycerol is its transformation in biogas (methane), through anaerobic fermentation [4]. Many microorganisms can use glycerol as a carbon source for growing. The presence of high concentration of chloride ions in the glycerol may inhibit the methanogenesis. Hence, dilution of the glycerol or co-feeding other wastewater streams is necessary.

Another approach to circumvent the high chloride concentration is to gradually replace pure glycerol with crude glycerol of biodiesel production. Thus, the microorganism can be acclimated to the extreme environment. For instance, a laboratory-scale reactor was fed with crude glycerol and nutrients, gradually decreasing the dilution of crude glycerol with respect of pure glycerol from 1:1500 to 1:5 (wt/wt). Around 97.5% of the organic matter in the effluent was converted without problems of instability [4].

A maximum theoretical production of 0.4 m^3 of methane per kg of glycerol can be obtained considering the stoichiometry of the reaction. Nevertheless, in practice, the amount is reduced to 50–70% of the theoretical value due to the presence of toxic substances.

The chemical transformation of glycerol appears as the most promising application, adding value to the biodiesel production chain. The world's energy matrix will be less dependent on fossil sources. Besides fuels, chemical products will also rely on renewable sources, and glycerol appears as an extremely good alternative. Being a by-product of biodiesel production, it does not compete with the fuel industry as ethanol does. The price of glycerol will be mostly dictated by the demand of biodiesel; as the use of this biofuel expands all over the world, the surplus of glycerol will be larger, forcing a decrease in the price. Figure 2.6 shows some possible products that can be obtained from glycerol, and there are many previous reviews on this subject in the literature [5–12]. Many of these processes will be discussed in more details in Chaps. 3, 4, and 5.

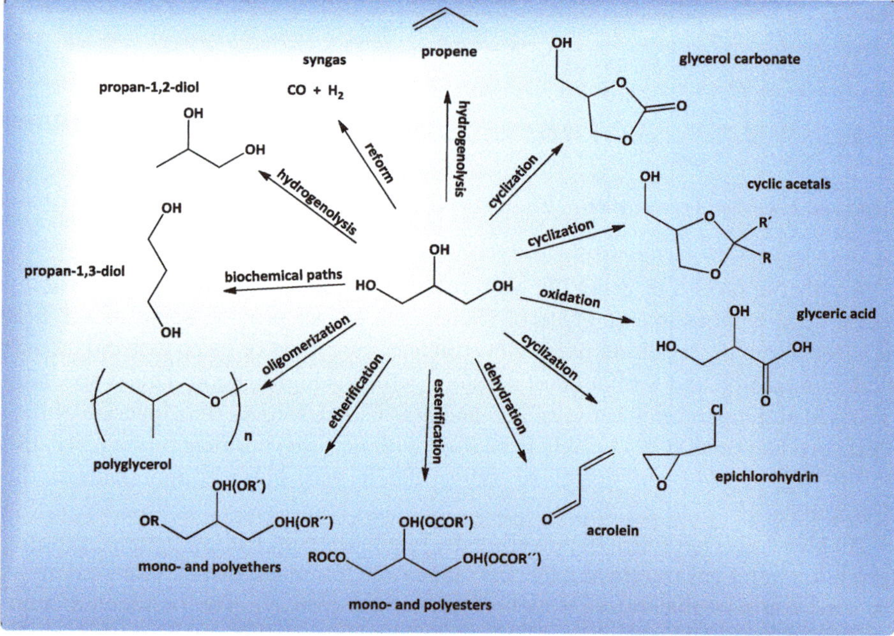

Fig. 2.6 Some products that can be obtained from glycerol

2.3 Glycerol Purification

The crude glycerol of biodiesel production presents different compositions depending on the raw material used. Hence, it is not fully suitable for many purposes, especially chemical transformations. In general, purification of crude glycerol of biodiesel production comprises three steps: firstly, there occurs the neutralization of the free fatty acids and dissolved basic catalyst; then, the excess of methanol is separated by distillation and returns to the production process; finally, there occurs the purification through vacuum distillation [13].

The most traditional process of glycerol purification is distillation at reduced pressure. A product with a content of glycerol of at least 99.5 wt% can be obtained by thin film distillation and meets the requirements of the US Pharmacopeia (USP). The boiling point of glycerol at 760 mmHg is 290°C. At this temperature, part of the glycerol is thermally decomposed. At 12.5 mmHg, glycerol boils at 179–180°C without significant decomposition. The residues of glycerol distillation may comprise polyglycerol, fatty acids, resinous materials and ashes, which are mainly composed of sodium chloride formed upon the neutralization of the basic catalyst.

The crude glycerol phase can also be purified by ion exchange resins that remove the sodium from glycerol/water solutions of high salt concentration [14, 15]. Adsorption and membrane technologies can also be used. Many purification methods are based on the distillation of the glycerol phase to strip alcohol contaminants from glycerol [16].

Ion exchange purification is not economically viable for high concentrations of salts. Distillation and membrane technologies are commonly used in these cases to obtain ultrapure glycerol. Membrane technologies are more cost-effective than distillation, provided that some prior purification to reduce salts and organic matter has been taken place [17].

Purification of crude glycerol of biodiesel production can also be accomplished with phosphoric acid, obtaining a product with a final purity of over 86% [18, 19]. Potassium phosphate obtained as by-product could be potentially used as fertilizer. Glycerol of high purity can be obtained upon the extraction of crude glycerol with ethanol [20]. Salts and fatty acids can be separated through filtration and decantation, respectively.

Although vacuum distillation is a high-energy demand process, it is still the main process of purification of the crude glycerol. Alternative methods, such as treatment with phosphoric acid and membrane separation [21], can also be used. The main advantages of this later method are the low energy consumption, high mass transfer rate, operation at room temperature and it does not require solvents or additives. Nevertheless, the lifetime of membranes, as well as their price, still needs to be optimized, which involves additional research and development to make them economically and operationally feasible on a large scale.

There are three basic types of refined glycerine, based on their purity and potential end use: (a) technical grade, (b) US Pharmacopeia (USP) (Fig. 2.7) and (c) Food Chemicals Codex (FCC).

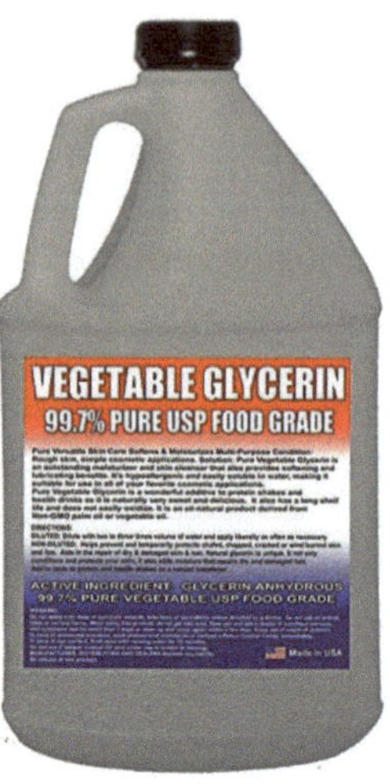

Fig. 2.7 USP grade glycerine for pharmaceutical or food applications (Source: https://br.pinterest.com/pin/371195194268102781)

Table 2.2 General specification of glycerol

Types of glycerol	Glycerol content (%)	Applications
Crude glycerol	70–90	Use as animal food supplement
Technical grade	99.5 Technical grade (not certified, mostly >96.0%)	Suitable as raw material for the chemical industry
US Pharmacopeia (USP)	99.5 USP (tallow based)	Used in cosmetic, pharmaceutical and food
US Pharmacopeia (USP)	99.5 USP (vegetable based)	Used in cosmetic, pharmaceutical and food
Food Chemical Codex (FCC)	99.7 USP/FCC-Kosher	Used in food

Technical grade glycerol is used as a chemical building block. Glycerine derived from vegetable oils and animal fats is suitable for pharmaceuticals (USP) and food products, whereas glycerine derived from vegetable oil is suitable for use in food (FCC). Table 2.2 summarizes the types and main applications of the different glycerol grades.

References

1. Nanda MR, Yuan Z, Qin W, Poirier MA, Chunbao X (2014) Purification of crude glycerol using acidification: effects of acid types and product characterization. Austin J Chem Eng 1:1–7
2. Fisher LJ, Erfle JD, Lodge GA, Sauer FD (1973) Effects of propylene glycol or glycerol supplementation of the diet of dairy cows on feed intake, milk yield and composition, and incidence of ketosis. Can J Anim Sci 53:289–296
3. Barbosa L C F, Moczydlower P, Pegoraro R T, Quintella C M A L T M H Process of enhanced oil recovery using by-products of biodiesel fabrication. BR 2009 PI 01604 (Brazil)

4. Viana MB, Freitas AV, Leitao RC, Pinto GAS, Santaella ST (2012) Anaerobic digestion of crude glycerol: a review. Environ Technol 1:81–92
5. Beatriz A, Araújo YJK, Lima DP (2011) Glycerol: a brief historic and application in stereoselective synthesis. Quim Nova 34:306–319
6. Behr A, Eilting J, Irawadi K, Leschinski J, Lindner F (2008) Improved utilization of renewable resources: new important derivatives of glycerol. Green Chem 10:13–30
7. Huber GW, Iborra S, Corma A (2006) Synthesis of transportation fuels from biomass: chemistry, catalysts and engineering. Chem Rev 106:4044–4098
8. Jérôme F, Pouilloux Y, Barrault J (2008) Rational design of solid catalysts for the selective use of glycerol as a natural organic building block. ChemSusChem 1:586–613
9. Mota CJA, Silva CXA, Gonçalves VL (2009) Glycerochemistry: new products and processes from glycerin of biodiesel production. Quim Nova 32:639–648
10. Pagliaro M, Ciriminna R, Kimura H, Rossi M, Pina CD (2007) From glycerol to value-added products. Angew Chem Int Ed 46:4434–4440
11. Zeng Y, Chen X, Shen Y (2008) Commodity chemicals derived from glycerol, an important biorefinery feedstock. Chem Rev 108:5253–5277
12. Zhou C, Beltramini JN, Fan YX, Lu GQ (2008) Chemoselective catalytic conversion of glycerol as a biorenewable source to valuable commodity chemicals. Chem Soc Rev 37:527–549
13. Ardi MS, Aroua MK, Awanis Hashim N (2015) Progress, prospect and challenges in glycerol purification process: a review. Renew Sustain Energy Rev 42:1164–1173
14. Carmona M, Valverde J, Pérez A (2008) Purification of glycerol/water solutions from biodiesel synthesis by ion exchange: sodium removal part I. J Chem Technol Biotechnol 84:738–744
15. Carmona M, Lech A, de Lucas A, Perez A, Rodrigues JF (2009) Purification of glycerol/water solutions from biodiesel synthesis by ion exchange: sodium and chloride removal part II. J Chem Technol Biotech 84:1130–1135
16. Potthast R, Chung C, Mathur I, Johann Haltermann Ltd (2010) Purification of glycerin obtained as a bioproduct from the transesterification of triglycerides in the synthesis of biofuel. United States patent application 12290728
17. Manosak R, Limpattayanate S, Hunsom M (2011) Sequential-refining of crude glycerol derived from waste used-oil methyl ester plant via a combined process of chemical and adsorption. Fuel Process Technol 92:92–99
18. Hajek M, Skopal F (2010) Treatment of glycerol phase formed by biodiesel production. Bioresour Technol 101:3242–3245
19. Javani A, Hasheminejad M, Tahvildari K, Tabatabaei M (2012) High quality potassium phosphate production through step-by-step glycerol purification: a strategy to economize biodiesel production. Bioresour Technol 104:788–790
20. Kongjao S, Damronglerd S, Hunsom M (2010) Purification of crude glycerol derived from waste used-oil methyl ester plant. Korean J Chem Eng 27:944–949
21. Strathmann H (1981) Membrane separation processes. J Membr Sci 9:121–189

Chapter 3
Biotechnological Routes of Glycerol Transformation in Valuable Chemicals

Abstract Biotechnological transformations of glycerol from the biodiesel industry are highlighted in this chapter. Several strategies based on the biochemical transformation have been studied to convert the residual glycerol into products of higher-added value, such as 1,3-propanediol, 3-hydroxypropanal, ethanol, 2,3-butanediol, lactic acid, succinic acid, citric acid, dihydroxyacetone and hydrogen. The biotechnological conversion of glycerol is an interesting alternative, because the processes are usually highly selective to the desired product and operated near ambient temperature and pressure. This chapter emphasizes the advantages and disadvantages of the biotechnological transformation of glycerol, as well as the factors that still need to be improved for a proper industrial application.

Keywords Biotechnology • Glycerol • 1,3-Propanediol • Ethanol • Lactic acid • Succinic acid • Citric acid • Dihydroxyacetone • Hydrogen

3.1 Introduction

Biochemical processes have some advantages over traditional thermochemical ones. They usually require mild operating conditions, involve less toxic reagents and are selective to the desired product. Enzymes are at the centre of the biotechnological transformations, being normally specific for reactants and products. In addition, the process is conveniently carried out with addition of nutrients and microorganisms [1].

The chemical transformations that occur in living systems are catalysed by different enzymes that specifically convert the substrates into products. The application of biotechnological processes has been known for long time. Nevertheless, due to economic factors, they have become important and attractive in more recent years [2]. Enzymes normally work according to the lock and key principle, in which a specific substrate interacts with the active site making the process highly selective. In addition, the reactions are usually carried out near ambient temperature and pressure; some enzymes still present high stability in organic solvents [3].

The use of biotechnological processes in the chemical industry has considerably increased in the past 10 years. They represented 3% of the technologies in 2004 but accounted for 15% of the processes in 2015 (Fig. 3.1). Biotechnology is still mostly

C.J.A. Mota et al., *Glycerol*, DOI 10.1007/978-3-319-59375-3_3

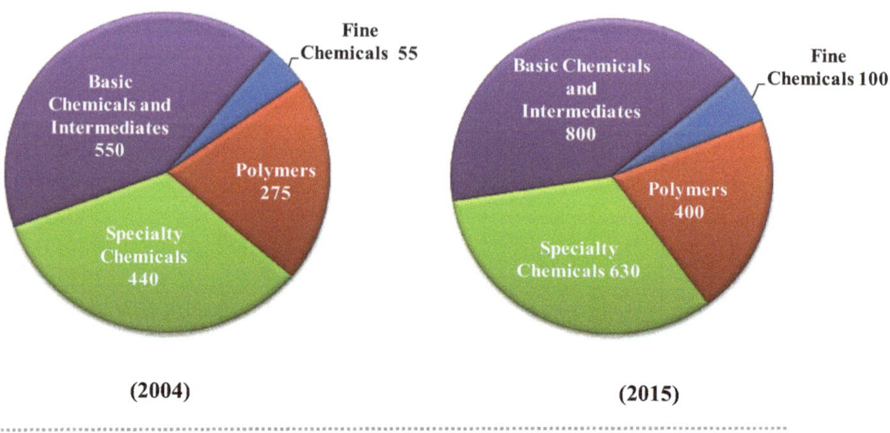

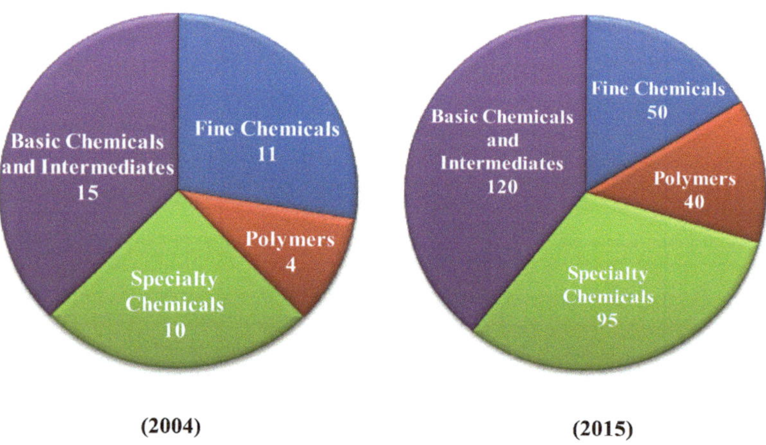

Fig. 3.1 Share of biotechnological and chemical processes, in volume of sales, in 2004 and in 2015 (Adapted from Ref. [4])

concentrated in the production of fine chemicals and specialty, but there is a continuous growth of biotechnological processes directed to basic chemicals, due to environmental aspects, safety and use of renewable feedstock. The global sales of chemical products in 2004 were 1.32 trillion euros, increasing to 1.93 trillion euros in 2015. In relation to the biotechnological processes, the total sales accounted for 40 billion euros in 2004 but showed a significant increase in 2015, reaching 305 billion euros [5].

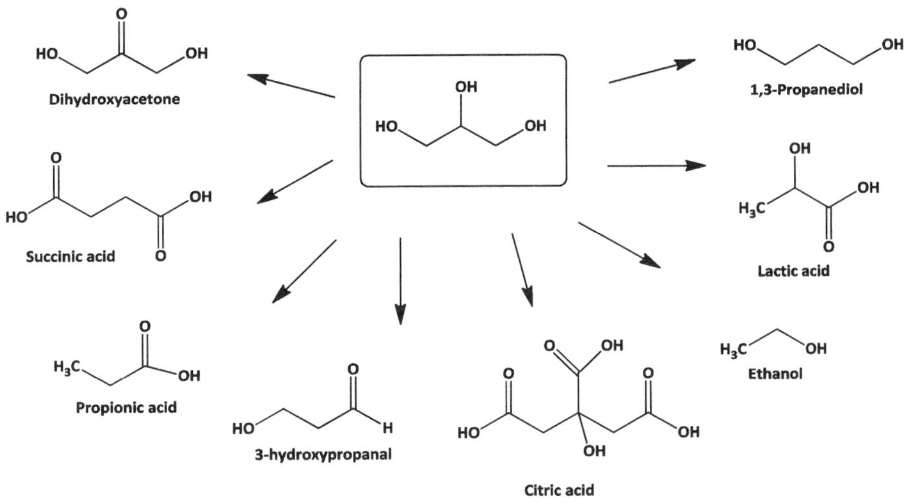

Fig. 3.2 Products obtained from glycerol through biotechnological routes

The biochemical conversion of glycerol presents great advantages due to the high level of reduction of the glycerol molecule compared to conventional feedstocks. In addition, the conversion of glycerol into phosphoenolpyruvate or pyruvate has advantages in comparison with glucose or xylose metabolism, requiring less energy demand for subsequent conversions [4].

Glycerol can be metabolized by many bacteria and yeasts under aerobic and anaerobic conditions. In general, the assimilation of glycerol by microorganisms involves active and passive transport across the plasma membrane. Figure 3.2 shows the different products that could be obtained from glycerol using the biochemical route [6]. After passing through the plasma membrane, glycerol can be catabolized by several independent metabolic routes, as shown in Scheme 3.1. Dehydration yields 3-hydroxypropanal; oxidation produces dihydroxyacetone, which ultimately leads to pyruvate-derived bioproducts.

3.2 1,3-Propanediol

1,3-Propanediol is used as a monomer for the synthesis of polyethers, polyurethanes and polyesters. Polytrimethylene terephthalate (PTT) is a synthetic fibre used in the production of carpets. It is obtained by the reaction of the diol with terephthalic acid. 1,3-Propanediol may also find applications in the cosmetic, food, pharmaceutical and textile sectors, as well as in the production of adhesives, solvents, detergents and resins. Although the production of 1,3-propanediol from petroleum derivatives is known, the biotechnological pathway has been considered environmentally friend and highly selective. For these reasons, this route has gained attention in recent years [6, 7].

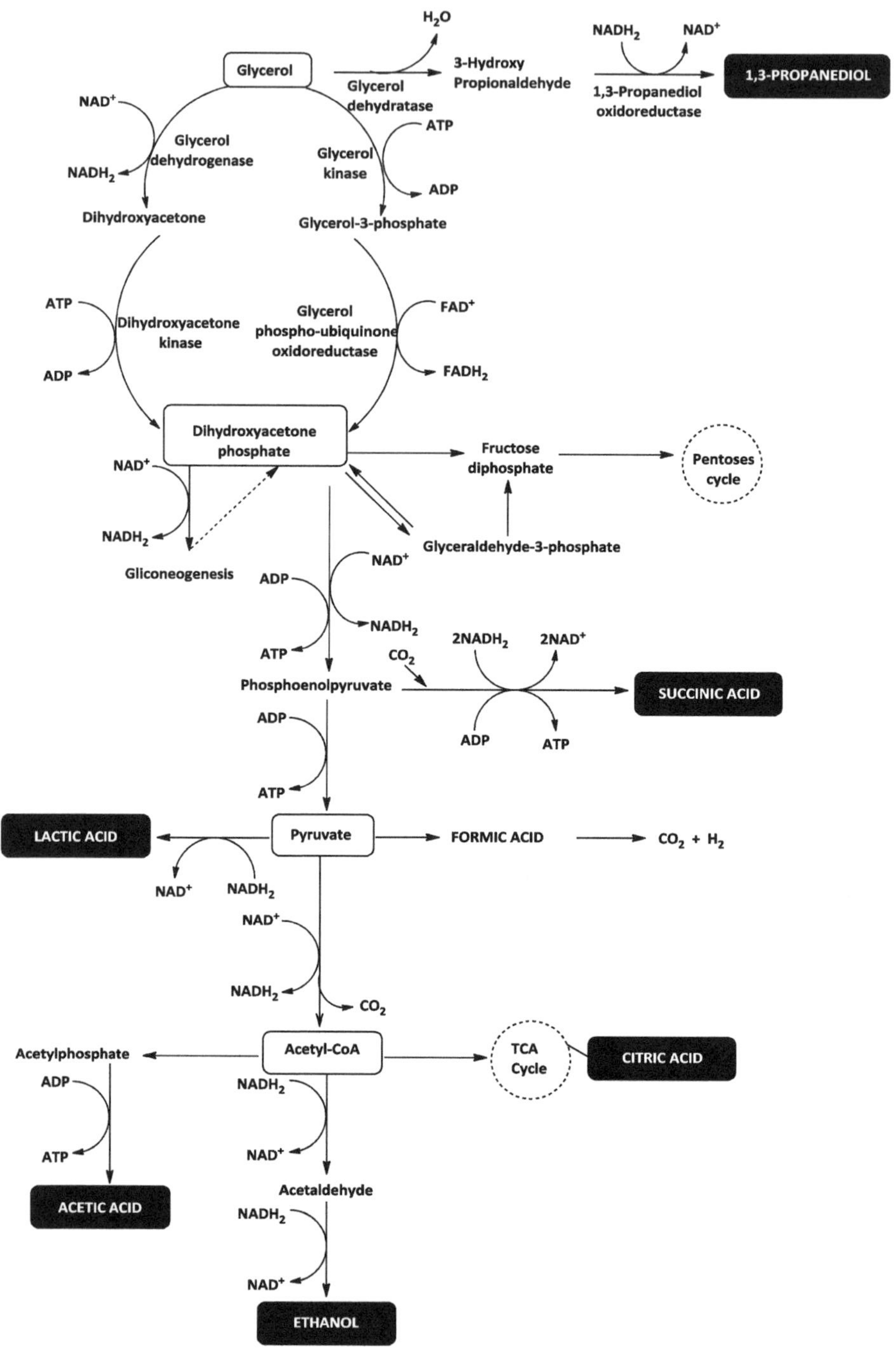

Scheme 3.1 Metabolic pathways of glycerol biochemical conversion and some possible products

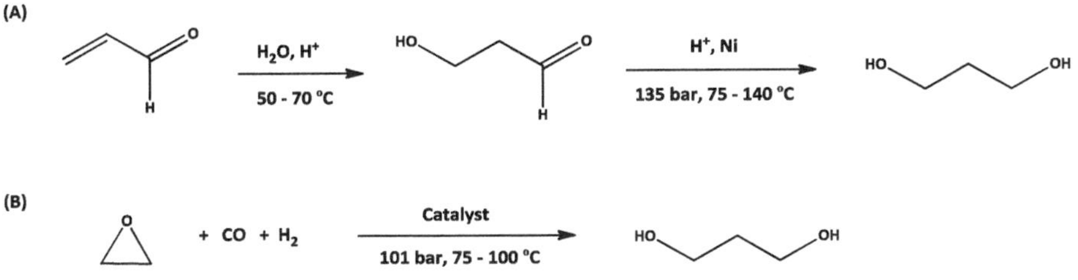

(A)

H_2O, H^+ / 50 - 70 °C ; H^+, Ni / 135 bar, 75 - 140 °C

(B)

+ CO + H_2 ; Catalyst / 101 bar, 75 - 100 °C

Scheme 3.2 Thermochemical synthesis of 1,3-propanediol

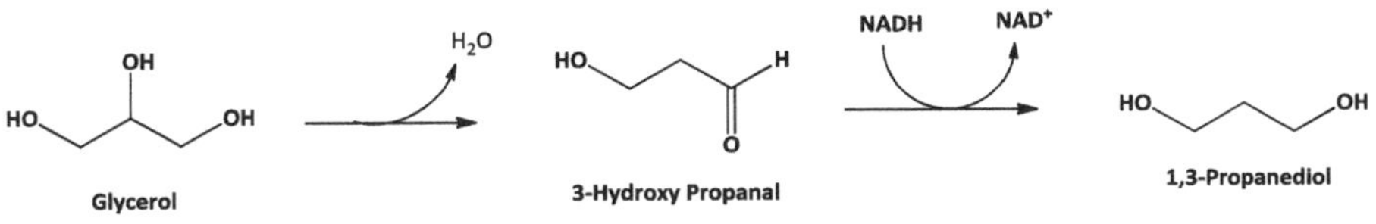

Glycerol → (H_2O) → 3-Hydroxy Propanal → (NADH → NAD^+) → 1,3-Propanediol

Scheme 3.3 Biochemical pathway of glycerol transformation to 1,3-propanediol

There are two chemical routes for the production of 1,3-propanediol. One of them (Fig. 3.2a) uses acrolein (2-propenal) as raw material. Hydration to 3-hydroxypropionic acid and subsequent hydrogenation in the presence of a catalyst yields 1,3-propanediol. The other route (Scheme 3.2b) involves the hydroformylation of ethylene oxide under high pressure and in the presence of a catalyst. However, these methods are expensive, involve toxic raw materials or intermediates and depend on oil.

The world's leading producer of 1,3-propanediol is the American chemical company DuPont, which used a biotechnological process with genetically modified *Escherichia coli* and corn as the main substrate. From a biochemical point of view, glycerol is considered one of the best natural substrates for the production of 1,3-propanediol. The biochemical pathway involves the dehydration of glycerol to 3-hydroxypropanal, which is catalysed by glycerol dehydratase enzyme. In the sequence, there occurs the reduction of the carbonyl group by glycerol dehydrogenase to yield 1,3-propanediol (Scheme 3.3) [8, 9].

Several microorganisms possess the ability to convert glycerol into 1,3-propanediol. Among them are *Klebsiella pneumoniae*, *Clostridia butyricum*, *Clostridium pasteurianum*, *Enterobacter agglomerans*, *Enterobacter aerogenes*, *Citrobacter freundii* and *Lactobacillus reuteri*. All of them are able to grow anaerobically with glycerol as a source of carbon and energy [10]. The bioproduction of 1,3-propanediol from glycerol can be made by various bacterial strains. There are several studies in the literature on the effects of process variables, such as pH, temperature and substrate feed control. Table 3.1 shows some selected literature papers on the production of 1,3-propanediol from glycerol.

Crude glycerol of biodiesel production contains nutrients such as phosphorus, magnesium, calcium and nitrogen, which are used by the microorganisms during the fermentative processes. On the other hand, chloride may be a limiting factor for the growth of some microorganisms.

Table 3.1 Production of 1,3-propanediol from glycerol using various microorganisms

Microorganism	C_{max} (g L^{-1})[a]	Y (g/g)[b]	P (g L^{-1} h^{-1})[c]	References
C. butyricum	47.1	0.53	1.12	[11]
C. butyricum	93.7	0.52	3.30	[12]
C. butyricum	54.2	0.54	1.55	[13]
C. butyricum	76.2	0.51	2.30	[14]
C. butyricum	56	0.51	1.90	[15]
K. pneumoniae	20.4	0.51	2.92	[16]
K. pneumoniae	70	0.70[d]	0.97	[17]
K. pneumoniae	71.6	0.54	1.93	[18]
K. pneumoniae	24.0	0.46[d]	1.0	[19]
K. pneumoniae	12.2	0.70	1.53	[20]
K. oxytoca	12,6	0.36	0.39	[21]
K. oxytoca	13.8	0.45	1.19	[22]
L. reuteri	10.6	0.66	1.42	[23]
L. reuteri	28.7	0.78	1.05	[24]
L. reuteri	13	0.70	1.08	[25]
C. freundii	23.3	0.61	0.97	[26]
C. freundii	16.3	0.53	8.2	[27]
S. blattae	13.84	0.45	1.19	[28]

[a]Maximum concentration of 1,3-propanediol
[b]Yield of 1,3-propanediol
[c]Productivity of 1,3-propanediol
[d]Mol/mol

3.3 Ethanol

Bioethanol is an important biofuel used worldwide. In 2015, the world's production of ethanol accounted for more than 97 billion litres. Besides being used as biofuel, ethanol also finds applications in various industry sectors (Fig. 3.3). It is possible to produce ethanol from glycerol using microorganisms of the type *Citrobacter* spp., *Klebsiella* spp., *Enterobacter* spp. and *Escherichia* spp. under anaerobic conditions [29].

Although the yeast *Saccharomyces cerevisiae* is the most common microorganism used in sugar fermentation to produce ethanol, it is not capable of converting glycerol to this alcohol. In addition, the glycerol obtained from the biodiesel industry contains high levels of contaminants, in particular methanol, which is considered a cytotoxic substance. The use of effective microorganisms is of paramount importance, because of the relatively low process productivity. The presence of additional nutrients, such as nitrogen, calcium, magnesium, phosphorus and sodium in the glycerol residue, has a positive effect on the growth of microorganisms, whereas impurities such as methanol and free fatty acids can dramatically affect their growth [30].

Ethanol is not only used as a feedstock and gasoline additive but also as a reagent for the production of biodiesel. Thus, the possibility of generating ethanol from glycerol is of interest to the biodiesel industry.

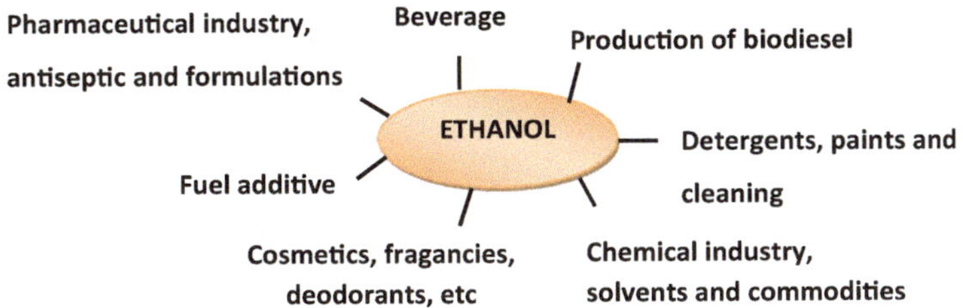

Fig. 3.3 Main ethanol usages

Ethanol can be obtained from ethylene or by fermentation of carbohydrates. The cost of producing ethanol is dependent on the raw material and operational expenditure. The production of ethanol from glycerol could reduce the cost of production by approximately 40%. In addition, when dealing with glycerol from biodiesel, the use of several microorganisms seems to be more advantageous than working with pure cultures, because of the contaminants present in the glycerol phase [31].

Presently, ethanol is produced from sugarcane, sugar beet, corn and cellulosic raw materials. Although many microorganisms have been potentially identified to produce ethanol from glycerol, the process still needs to be optimized, due to prolonged times, which could reach 42 days, and low yields of product [32, 33].

The biochemical pathway is complex and involves the transformation of glycerol in glyceraldehyde 3-phosphate, which then goes to pyruvate before ending in ethanol after a series of biochemical transformations.

3.4 Lactic Acid

Lactic acid (2-hydroxypropanoic acid) is normally produced from sugar fermentation by lactic bacteria. The use of lactic acid is present in several industrial branches, such as in the food industry, where it acts as a preservative and emulsifiers. In the pharmaceutical industry, lactic acid is employed in the preparation of lotions and cosmetics. It also finds applications in the chemical industry as a building block for the production of acrylic acid and propylene glycol [34].

Lactic acid is also used in the production of the polylactate polymer (PLA), which is biodegradable and can be used in medicine. In the chemical industry, PLA is employed in the manufacture of biodegradable plastics (Fig. 3.4). In the food industry, lactic acid is used as pH lowering, antimicrobial agent, stabilizer, humectant, emulsifier and plasticizer [35].

Lactic acid can be obtained, either, by the fermentative action of bacteria, fungi and yeasts or by chemical synthesis. However, the fermentative pathway has advantages. Sugars are the best sources of carbon for the bacteria used to produce lactic acid. There is also the need of a source of nitrogen, vitamins and minerals [36].

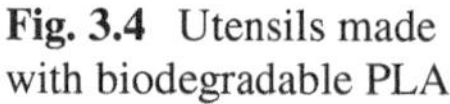
Fig. 3.4 Utensils made with biodegradable PLA

The biochemical pathway of glycerol to lactic acid involves the formation of dihydroxyacetone phosphate, which is considered an important intermediate in the biochemical pathway of glycerol conversion to some compounds of commercial interest (Scheme 3.1) [37].

3.5 Citric Acid

The amount of citric acid (2-hydroxypropane-1,2,3-tricarboxylic acid) produced annually in the world is greater than the production of any other organic acid obtained by fermentative processes. This market increases at a rate of 5% per year, especially because of the food and beverage sectors. Citric acid shares 50–65% of the food and beverage acidulants market, against 20–25% of phosphoric and 5% malic acid. In the pharmaceutical industry, citric acid is used as an effervescent agent [38].

For many years, it was believed that *Aspergillus niger* was the only fungus capable of producing citric acid in large quantities, through the fermentation of molasses, beet, sugar cane, sucrose or glucose. However, in 1960 it was shown that some yeast may also be used to synthesize citric acid, such as *Candida guilliermondii*, *Candida lipolytica* and *Y. lipolytica*. This later yeast is highly tolerant to the presence of metallic ions, allowing its use in less refined substrates. Since the last decade, glycerol has become an alternative substrate for the synthesis of citric acid [39].

A large number of microorganisms, including bacteria, fungi and yeasts, have been used to produce citric acid. Nevertheless, the yields are low. *A. niger* is still used for commercial production. This microorganism is easy to handle, has the ability to ferment a variety of cheap raw materials and shows high yield of citric acid.

The biotechnological production of citric acid is strongly influenced by the composition of the medium, carbon source concentration, nitrogen and phosphate limitation, pH and morphology of the producing microorganism, among others. Certain nutrients need to be in excess (such as sugars), others at limiting levels (such as nitrogen and phosphate) and others below the threshold values (trace metals, particularly manganese).

About 99% of world's production of citric acid occurs via microbial processes. About 70% of the total production of 1.5 million tons per year is used in the food and beverage industry, as an acidifier or antioxidant to preserve or enhance the flavours and aromas of fruit juices, ice cream and marmalades. Around 20% is used, as such, in the pharmaceutical industry as antioxidant to preserve vitamins, or as an effervescent agent, pH corrector, or in the form of iron citrate, as a source of iron for the body. In the chemical industry, which responds for 10% of the citric acid market, it is employed as a foaming agent for the softening and treatment of textiles.

3.6 Succinic Acid

Succinic acid (or butanedioic acid) is a dicarboxylic acid that can be used as a precursor to many chemicals of industrial importance, including adipic acid, 1,4-butanediol, tetrahydrofuran, *N*-pyrrolidinone, 2-pyrrolidinone and others. Among the applications of succinic acid are synthetic resins, biodegradable polymers, chemical intermediates and additives (Fig. 3.5).

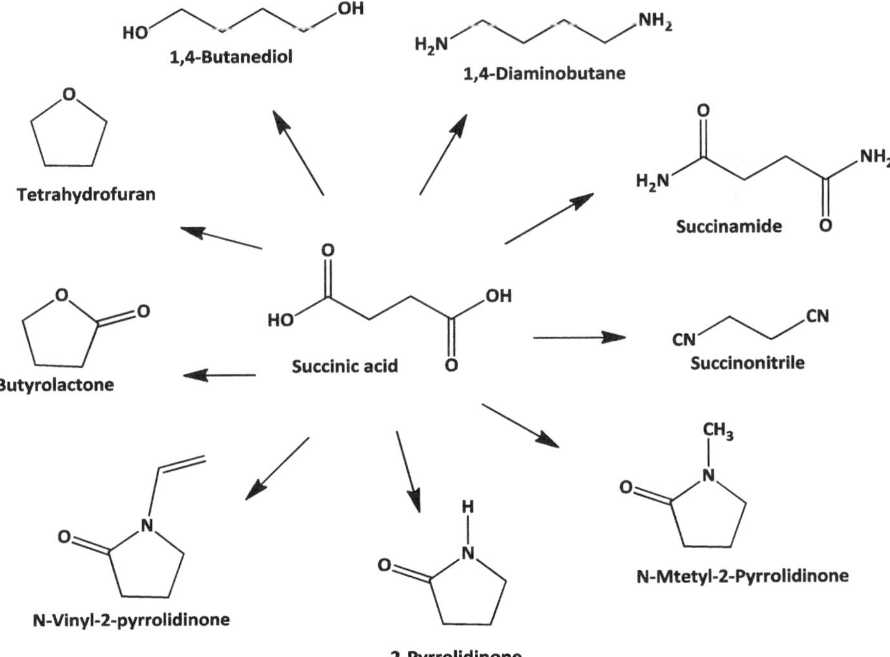

Fig. 3.5 Some chemicals produced from succinic acid

Presently, a major part of succinic acid is produced from fossil sources, through the oxidation of *n*-butane to maleic anhydride, followed by hydration and hydrogenation. The cost of maleic anhydride is what most significantly contributes to the overall costs of succinic acid from this route. Recent developments have focused on biotechnological alternatives for the production of succinic acid [40, 41].

Using the biotechnological route of succinic acid production, many different microorganisms have been screened. Among them are *Anaerobiospirillum succiniciproducens* and *Actinobacillus succinogenes*, because of their ability to produce a relatively large amount of succinic acid. However, the formation of by-products such as acetic, formic and lactic acid is common and reduces the yield and productivity of succinic acid, besides increasing the complexity and cost of the process.

The biotechnological production of succinic acid on an industrial scale has been developed by chemical companies such as BASF, BioAmber and DSM, using sugars as feedstock. Due to availability and low price of glycerol, it could be used in the production of succinic acid. *Actinobacillus succinogenes* is one of the best microorganisms to transform glycerol into succinic acid. However, there is a redox imbalance during cell growth that can be solved using an external electron acceptor such as DMSO (dimethyl sulfoxide). Although high yields of succinic acid have been reported from glycerol, there are still several obstacles that impair the economic feasibility of the route.

Figure 3.6 illustrates the biochemical conversion of glycerol through the glycolytic intermediates, phosphoenolpyruvate (PEP) or pyruvate (PYR). The fermentative metabolism allows higher productivity from glycerol compared to glucose or xylose [42].

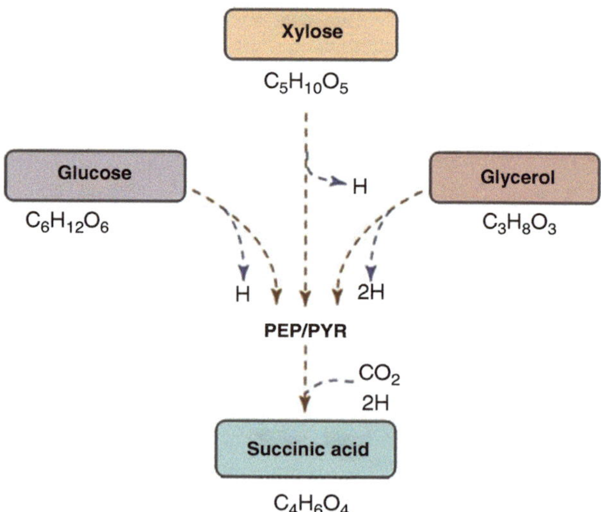

Fig. 3.6 Biochemical conversion of glycerol, glucose and xylose into the glycolytic intermediates to yield succinic acid

3.7 Propionic Acid

Propionic acid, also known as propanoic acid, has broad industrial applications in pharmaceuticals, herbicides, cosmetics, additives in cellulosic plastics and food preservatives, as well as antifungal agent. Presently, the industrial production of propionic acid relies on fossil-based resources. It is mainly produced from ethylene, carbon monoxide and steam (Reppe process) or from ethanol and carbon monoxide using boron trifluoride as catalyst (Larson process). Another route involves the oxidation of propionaldehyde, which can be obtained as by-product in Fischer–Tropsch processes [43].

The biotechnological route of propionic acid production through fermentation normally uses *Propionibacterium* sp. This microorganism is capable of producing propionic acid from various sources of cheap raw materials, such as glycerol from biodiesel production [44]. Notwithstanding, propionic acid from fermentation is still more expensive than the one obtained from chemical route [45].

The biotechnological production of propionic acid from carbohydrate-based feedstocks, including glucose and whey lactose, has long been studied. However, there are few studies on the use of glycerol as a source of carbon in the fermentation to produce propionic acid. Glycerol is in a reduced state, compared with sugars, which favours the production of metabolites. On the other hand, it can also undergo redox imbalance in metabolism and, thus, inhibits cell growth. In addition, the conventional fermentation process exhibits a low yield of propionic acid due to strong inhibition [46].

The use of microorganisms resistant to inhibitory components, such as salts and organic components present in the crude glycerol, is still a challenge. The use of glycerol as the sole carbon source in the propionic acid production process results in a lower product build-up, compared to glucose, xylose, maltose and sucrose. However, an increase in crude glycerol concentration from 20 g/L to 40 g/L in the culture medium did not lead to an enhanced production of propionic acid (11.16 g/L versus 9.37 g/L, respectively), because of the concomitant increase of inhibitors present in the raw material [47].

3.8 Dihydroxyacetone

Dihydroxyacetone (DHA) is one of the most commercially important chemicals used in cosmetics, medicine and pharmaceuticals. It is also used in the food sector and as raw material in the fine chemical industry. It is the main component of artificial tan or sunless tan products. Upon application on the skin, DHA reacts with the amino acids through the Maillard reaction producing melanoidins, which cause the appearance of tan (Fig. 3.7).

Dihydroxyacetone can be produced via the catalytic oxidation of glycerol, which is discussed in Chap. 5. However, this pathway usually involves expensive catalysts and produces several by-products. The microbial fermentation route, on the other

Fig. 3.7 Commercial self-tanner product based on DHA (Source: https://br.pinterest.com/pin/377528381244873219/)

hand, is more economic and ecological efficient. The reaction occurs with the aid of the glycerol dehydrogenase enzyme, with concomitant transformation of the NAD in NADH (Scheme 3.4) [48].

The higher price of dihydroxyacetone (US$150/kg) relative to that of pure glycerol (US$0.60/kg) makes the conversion particularly attractive. On the other hand, the purification and crystallization of the final product is complicated, laborious, time-consuming and increases the production costs. The immobilization of enzymes using nanostructured materials has been shown to be efficient for prolonged operations, whilst inhibiting denaturation and leaching [49].

The production of dihydroxyacetone from microbial route is more efficient than from chemical conversion. In addition, biotechnological routes are more acceptable to the pharmaceutical and chemical industries, because of stringent environmental constraints and the problems involved with chemical synthesis methods [50].

The use of microorganisms such as *Gluconobacter oxydans* results in a final product with few impurities. Thus, extraction, purification and crystallization of the product can be carried out more effectively, leading to a dramatic reduction of the process costs. *Gluconobacter oxydans* has been widely used in commercial operations, because of the high selectivity towards dihydroxyacetone (around 75%) as well as the stability and formation of less by-products [51–54].

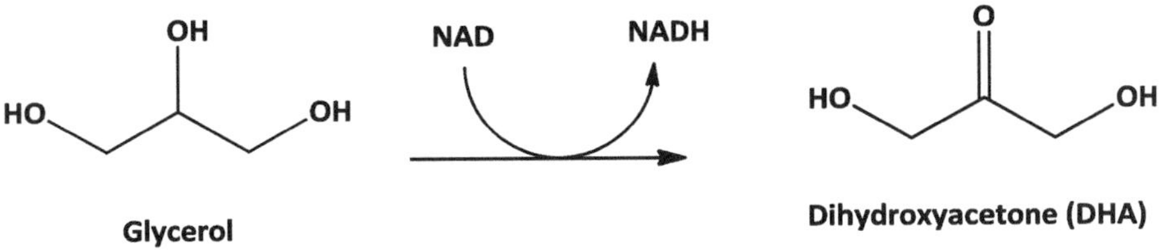

Scheme 3.4 Biochemical transformation of glycerol in DHA

3.9 Hydrogen

Hydrogen has many applications in the chemical industry. Approximately half of all the hydrogen produced in the world is used in the synthesis of ammonia, which, in turn, is important in the manufacture of fertilizers. About 37% of the hydrogen produced is used in oil refining, to remove impurities or to upgrade heavy oil into lighter fractions and higher-value products. Around 8% of the hydrogen market is driven to the production of methanol. The remaining 5% is used in a wide variety of chemical processes, metallurgy and others [55].

Hydrogen is considered a clean and green fuel. Nevertheless, 95% of the hydrogen produced nowadays comes from non-renewable sources. The use of hydrogen as fuel requires the use of an economically viable and renewable source. Generally, there are four basic processes for the production of hydrogen from non-fossil sources: water electrolysis, thermochemical processes, radiolytic processes and biological processes [56–58].

Hydrogen production by microorganisms occurs at ambient temperature and pressure, being less energy demanding than conventional technologies. Microorganisms produce hydrogen through two main pathways: photosynthesis and fermentation. Photosynthesis is a light-dependent process that can be carried out by algae, photosynthetic bacteria and cyanobacteria, whereas the fermentative hydrogen production is conducted by microorganisms [59].

The production of hydrogen by the fermentative biotechnological pathway presents some advantages, as it does not depend on light and may employ a variety of carbon sources, among them glycerol. In addition, microorganisms exhibit rapid growth and are unaffected by oxygen, unlike the photosynthetic process, where oxygen inhibits the hydrogenase enzyme that is responsible for H_2 production. Some microorganisms, such as methanogenic or sulfo-nuclear bacteria, are hydrogen consuming and should be avoided in the fermentative process [60].

The production of hydrogen through the anaerobic digestion of crude glycerol has gained prominence. Mixed cultures of microorganisms are more practical and robust than those using well-defined pure cultures. The use of glycerol as a carbon source promotes the reuse of waste materials, thus reducing the number of by-products stored in the industry. However, the presence of impurities, such as salts, methanol and soap, can inhibit the process by the metabolic pathway. Therefore, a pretreatment of the glycerol of biodiesel production is often necessary [61]. Certain microorganisms can grow anaerobically in glycerol as the sole source of carbon to generate hydrogen [62].

Bacteria capable of producing large-scale H_2 can be found naturally in soils, water and sewage. These materials can be used as inoculum in media for fermentative H_2 production. A variety of pure or mixed cultures of bacteria have been used for the production of H_2 using glycerol as a substrate. Among them are *Escherichia*, *Pseudomonas*, *Thermococcus*, *Ruminococcus*, *Hafnia*, *Citrobacter* and *Ethanoligenense*. The *Thermotoga bacterium* is also studied for the generation of hydrogen under thermophilic conditions of 80°C and pH of 7.0–7.4 [63].

References

1. Doble M, Gummadi SN (2007) Biochemical engineering. Eastern Economy Edition, ISBN-978-81-203-3052-8
2. Oliveira LG, Mantovani SM (2009) Biological transformations: contributions and perspectives. Quím Nova 32:742–756
3. Castro HF, Mendes AA, Santos JC, Aguiar CL (2004) Modification of oils and fats by biotransformation. Quím Nova 27:146–156
4. Komider A, Leja K, Czaczyk K (2011) Improved utilization of crude glycerol by-product from biodiesel production. In: Biodiesel-quality, emissions and by-product. Intech, Croatia, ISBN 978-953-307-784-0
5. Saxena RK, Anand P, Saran S, Isar J (2009) Microbial production of 1,3-propanediol: recent developments and emerging opportunities. Biotechnol Adv 27:895–913
6. Blaschek HP, Ezeji TC, Scheffran J (2010) Biofuels from agricultural wastes and byproducts. Wiley-Blackwell, Ames, IA, ISBN: 978-0-8138-0252-7
7. Vivek N, Pandey A, Binod P (2017) Production and applications of 1,3-propanediol. In: Current developments in biotechnology and bioengineering, p 719–738, ISBN 978-0-444-63662-1
8. Vivek N, Pandey A, Binod P (2016) Biological valorization of pure and crude glycerol into 1,3-propanediol using a novel isolate *Lactobacillus brevis* N1E9.3.3. Bioresour Technol 213:222–230
9. Raynaud C, Sarcabal P, Meynial-Salles I, Croux C, Soucaille P (2003) Molecular characterization of the 1,3-propanediol (1,3-PD) operon of *Clostridium butyricum*. Proc Natl Acad Sci 100:5010–5015
10. Pflügl S, Marx H, Mattanovich D, Sauer M (2012) 1,3-Propanediol production from glycerol with *Lactobacillus diolivorans*. Bioresour Technol 119:133–140
11. Papanikolaou S, Fakas S, Fick M, Chevalot I, Galiotou-Panayotou M, Komaitis M, Marc I, Aggelis G (2008) Biotechnological valorisation of raw glycerol discharged after bio-diesel (fatty acid methyl esters) manufacturing process: production of 1,3-propanediol, citric acid and single cell oil. Biomass Bioenergy 32:60–71
12. Wilkens E, Ringe AK, Hortig ID, Willke T, Vorlop KD (2012) High-level production of 1,3-propanediol from crude glycerol by *Clostridium butyricum* AKR102a. Appl Microbiol Biotechnol 93:1057–1063
13. Powałowska DS (2014) 1,3-Propanediol production from crude glycerol by *Clostridium butyricum* DSP1 in repeated batch. Electron J Biotechnol 17:322–328
14. Wilkens E, Ringel AK, Hortig D, Willke T, Vorlop KD (2012) High-level production of 1,3-propanediol from crude glycerol by *Clostridium butyricum* AKR102a. Appl Microbiol Biotechnol 93:1057–1763
15. Biebl H, Marten S, Hippe H, Deckwer WD (1992) Glycerol conversion to 1,3-propanediol by newly isolated clostridia. Appl Microbiol Biotechnol 36:592–597

16. Silva GP, Lima CJB, Contiero J (2015) Production and productivity of 1,3-propanediol from glycerol by *Klebsiella pneumoniae* GLC29. Catal Today 257:259–266
17. Ji XJ, Huang H, Zhu JG, Hu N, Li S (2009) Efficient 1,3-propanediol production by fed-batch culture of *Klebsiella pneumoniae*: the role of pH fluctuation. Appl Biochem Biotechnol 159:605–613
18. Jin P, Li S, Lu SG, Zhu JG, Huang H (2011) Improved 1,3-propanediol production with hemicellulosic hydrolysates (corn straw) as cosubstrate: impact of degradation products on *Klebsiella pneumoniae* growth and 1,3-propanediol fermentation. Bioresour Technol 102:1815–1182
19. Rossi DM, Souza EA, Flôres SH, Ayub MAZ (2013) Conversion of residual glycerol from biodiesel synthesis into 1,3-propanediol by a new strain of *Klebsiella pneumonia*. Renew Energy 55:404–409
20. Zhang GL, Ma BB, Xu XL, Li C, Wang L (2007) Fast conversion of glycerol to 1,3-propanediol by a new strain of *Klebsiella pneumonia*. Biochem Eng J 37:256–260
21. Metsoviti M, Paraskevaidi K, Koutinas A, Zeng AP, Papanikolaou S (2012) Production of 1,3-propanediol, 2,3-butanediol and ethanol by a newly isolated *Klebsiella oxytoca* strain growing on biodiesel-derived glycerol based media. Process Biochem 47:1872–1882
22. Wojtusik M, Rodríguez A, Ripoll V, Santos VE, García JL, Ochoa FG (2015) 1,3-Propanediol production by *Klebsiella oxytoca* NRRL-B199 from glycerol. Medium composition and operational conditions. Biotechnol Rep 6:100–107
23. Vieira PB, Kilikian BV, Bastos RV, Perpetuo EA, Nascimento CAO (2015) Process strategies for enhanced production of 1,3-propanediol by *Lactobacillus reuteri* using glycerol as a co-substrate. Biochem Eng J 94:30–38
24. Jimenez RB, Lopez-Martinez LX, Cruz-Medina J, Espinosa-de-los-Monteros JJ, Garcia-Galindo HS (2011) Effect of glucose on 1,3-propanediol production by *Lactobacillus reuteri*. Rev Mex Ing Quim 10:39–46
25. Tobajas M, Mohedano AF, Casas JA, Rodriguez JJ (2009) Unstructured kinetic model for reuterin and 1,3-propanediol production by *Lactobacillus reuteri* from glycerol/glucose cofermentation. J Chem Technol Biotechnol 84:675–680
26. Drozdzynska A, Pawlicka J, Kubiak P, Kosmider A, Pranke D, Olejnik-Schmidt A, Czaczyk K (2014) Conversion of glycerol to 1,3-propanediol by *Citrobacter freundii* and *Hafnia alvei*–newly isolated strains from the Enterobacteriaceae. New Biotechnol 31:402–410
27. Pflugmacher U, Gottschalk G (1994) Development of an immobilized cell reactor for the production of 1,3-propanediol by *Citrobacter freundii*. Appl Microbiol Biotechnol 41:313–316
28. Rodriguez A, Wojtusik M, Ripoll V, Santos VE, Ochoa FG (2016) 1,3-Propanediol production from glycerol with a novel biocatalyst *Shimwellia blattae* ATCC 33430: operational conditions and kinetics in batch cultivations. Bioresour Technol 200:830–837
29. Maru BT, López F, Kengen SWM, Constantí M, Medina F (2016) Dark fermentative hydrogen and ethanol production from biodiesel waste glycerol using a co-culture of *Escherichia coli* and *Enterobacter* sp. Fuel 186:375–384
30. Stepanov N, Efremenko E (2017) Immobilised cells of *Pachysolen tannophilus* yeast for ethanol production from crude glycerol. New Biotechnol 34:54–58
31. Vikromvarasiri N, Haosagul S, Boonyawanich S, Pisutpaisal N (2016) Microbial dynamics in ethanol fermentation from glycerol. Int J Hydrogen Energy 41:15667–15673
32. Hao J, Lin R, Zheng Z, Liu H, Liu D (2008) Isolation and characterization of microorganisms able to produce 1,3-propanediol under aerobic conditions. World J Microbiol Biotechnol 24:1731–1740
33. Jarvis GN, Moore ERB, Thiele JH (1997) Formate and ethanol are the major products of glycerol fermentation produced by a *Klebsiella planticola strain* isolated from red deer. J Appl Microbiol 83:166–174
34. Datta R, Henry M (2006) Lactic acid: recent advances in products, processes and technologies: a review. J Chem Technol Biotechnol 81:1119–1129
35. Nampoothiri KM, Nair NR, John RP (2010) An overview of the recent developments in polylactide (PLA) research. Bioresour Technol 101:8493–8501

36. Demirci A, Pometto AL, Lee B, Hinz PN (1998) Media evaluation of lactic acid repeated-batch fermentation with Lactobacillus plantarum and Lactobacillus casei subsp. rhamnosus. J Agric Food Chem 46:4771–4774
37. Murakami N, Oba M, Iwamoto M, Tashiro Y, Noguchi T, Bonkohara K, Abdel-Rahman MA, Zendo T, Shimoda M, Sakai K, Sonomoto K (2016) L-Lactic acid production from glycerol coupled with acetic acid metabolism by *Enterococcus faecalis* without carbon loss. J Biosci Bioeng 121:89–95
38. Amaral PFF, Ferreira TF, Fontes GC, Coelho MAZ (2009) Glycerol valorization: new biotechnological routes. Food Bioprod Process 87:179–186
39. Rywinska A, Juszczyk P, Wojtatowicz M, Robak M, Lazar Z, Tomaszewska L, Rymowicz W (2013) Glycerol as a promising substrate for *Yarrowia lipolytica* biotechnological applications. Biomass Bioenergy 48:148–166
40. Song H, Lee SY (2006) Production of succinic acid by bacterial fermentation. Enzym Microb Technol 39:352–361
41. Cheng KK, Zhao XB, Zeng J, Zhang JA (2012) Review: biotechnological production of succinic acid. Biofuels Bioprod Biorefin 6:302–318
42. Yazdani SS, Gonzalez R (2007) Anaerobic fermentation of glycerol: a path to economic viability for the biofuels industry. Curr Opin Biotechnol 18:213–219
43. Liu Z, Ge Y, Xu J, Gao C, Ma C, Xu P (2016) Efficient production of propionic acid through high density culture with recycling cells of *Propionibacterium acidipropionici*. Bioresour Technol 216:856–861
44. Stowers CC, Cox BM, Rodriguez BA (2014) Development of an industrializable fermentation process for propionic acid production. J Ind Microbiol Biotechnol 41:837–852
45. Wallenius J, Pahimanolis N, Zoppe J, Kilpeläinen P, Master E, Ilvesniemi H, Seppälä J, Eerikäinen T, Ojamo H (2015) Continuous propionic acid production with Propionibacterium acidipropionici immobilized in a novel xylan hydrogel matrix. Bioresour Technol 197:1–6
46. Zhang A, Yang ST (2009) Propionic acid production from glycerol by metabolically engineered *Propionibacterium acidipropionici*. Process Biochem 44:1346–1351
47. Kośmider A, Drożdżyńska A, Blaszka K, Leja K, Czaczyk K (2010) Propionic acid production by Propionibacterium freudenreichii ssp. shermanii using crude glycerol and whey lactose industrial wastes. Pol J Environ Stud 19:1249–1253
48. Dikshit PK, Moholkar VS (2016) Kinetic analysis of dihydroxyacetone production from crude glycerol by immobilized cells of *Gluconobacter oxydans* MTCC 904. Bioresour Technol 216:948–957
49. Kumar GS, Wee Y, Lee I, Sun HJ, Zhao X, Xia S, Kim S, Lee J, Wang P, Kim J (2015) Stabilized glycerol dehydrogenase for the conversion of glycerol to dihydroxyacetone. Chem Eng J 276:283–288
50. Liu YP, Sun Y, Tan C, Li H, Zheng XJ, Jin KQ, Wang G (2013) Efficient production of dihydroxyacetone from biodiesel-derived crude glycerol by newly isolated *Gluconobacter frateurii*. Bioresour Technol 142:384–389
51. Hu ZC, Zheng YG, Shen YC (2011) Use of glycerol for producing 1,3-dihydroxyacetone by *Gluconobacter oxydans* in an airlift bioreactor. Bioresour Technol 102:7177–7182
52. Ma L, Lu W, Xia Z, Wen J (2010) Enhancement of dihydroxyacetone production by a mutant of *Gluconobacter oxydans*. Biochem Eng J 49:61–67
53. Hekmat D, Bauer R, Fricke J (2003) Optimization of the microbial synthesis of dihydroxyacetone from glycerol with *Gluconobacter oxydans*. Bioprocess Biosyst Eng 26:109–116
54. Bauer R, Hekmat D (2006) Development of a transient segregated mathematical model of the semicontinuous microbial production process of dihydroxyacetone. Biotechnol Prog 22:278–284
55. Nath K, Das D (2005) Hydrogen production by Rhodobacter sphaeroides strain O.U.001 using spent media of *Enterobacter cloacae* strain DM11. Appl Microbiol Biotechnol 68:533–541
56. Lay J, Lee Y, Noike T (1999) Feasibility of biological hydrogen production from organic fraction of municipal solid waste. Water Res 33:2579–2586

57. Das D, Veziroglu TN (2001) Hydrogen production by biological processes: a survey of literature. Int J Hydrog Energy 26:13–28
58. Dunn S (2002) Hydrogen futures: toward a sustainable energy system. Int J Hydrogen Energy 27:235–264
59. Vardar-Schara G, Maeda T, Wood TK (2008) Metabolically engineered bacteria for producing hydrogen via fermentation. Microb Biotechnol 1:107–125
60. Hawkes FR, Dinsdale R, Hawkes DL, Hussy I (2002) Sustainable fermentative hydrogen production: challenges for process optimization. Int J Hydrogen Energy 27:1339–1347
61. Sarma SJ, Brar SK, Sydney EB, Bihan LE, Buelna YG, Soccol CR (2012) Microbial hydrogen production by bioconversion of crude glycerol: a review. Int J Hydrogen Energy 37:6473–6490
62. Li C, Fang H (2007) Fermentative hydrogen production from wastewater and solid wastes by mixed cultures. Crit Rev Environ Sci Technol 37:1–39
63. Poleto L, Souza P, Eva MF, Beal LL, Torres RAP, Sousa MP, Laurino JP (2016) Selection and identification of microorganisms present in the treatment of waste water and activated sludge to produce biohydrogen from glycerol. Int J Hydrogen Energy 41:4374–4381

Chapter 4
Thermochemical Routes of Glycerol Transformation to Commodity Chemicals

Abstract The chemical industry ultimately produces polymers for the use of the general public. Today, most of the chemicals are based on oil, coal or natural gas, which reserves are limited and exploitation associated with global warming. In the past years, there were many developments of technological routes to produce chemicals from renewable. The possibility of replacing oil and gas by biomass puts the chemical industry at the forefront of the research and development of new processes. This chapter highlights recent advances in the conversion of glycerol to commodity chemicals, mostly used to produce polymers. Glycerol can be an interesting raw material to replace propylene-based chemicals, such as propylene glycol, acrolein and acrylic acid. Other developments are also discussed, such as a pioneer process of selective glycerol hydrogenolysis to propene itself, as well as production of syngas, hydrogen and methanol.

Keywords Chemical industry • Hydrogenolysis • 1,2-Propanediol • Propene • Acrolein • Acrylic acid • Epichlorohydrin • Methanol • Reforming • Commodities

4.1 The Chemical Industry

In economy, the term commodity refers to a marketable item of uniform quality and characteristics, independent of the producer or source.

Commodity chemicals may be defined as a group of substances, normally produced in large scale, that meet standard quality and purity characteristics, regardless of the producer or raw material. The major petrochemicals, such as ethylene, propylene, styrene, phenol and most of the common polymers, among others, can be considered commodities. The chemical industry is expected to grow at an annual rate around 3.6% in the next years. In 2010, the total sales of chemical products accounted for over 2.3 trillion euros, being one of the major industrial sectors in the world.

The petrochemical industry relies on the use of oil and gas derivatives and involves three generations (Fig. 4.1). The first generation is based on the production of light olefins (ethylene, propylene) and aromatics (benzene, toluene and xylenes). The major raw material for the production of olefins and aromatics is naphtha, a hydrocarbon fraction obtained from the atmospheric distillation of the petroleum,

C.J.A. Mota et al., *Glycerol*, DOI 10.1007/978-3-319-59375-3_4

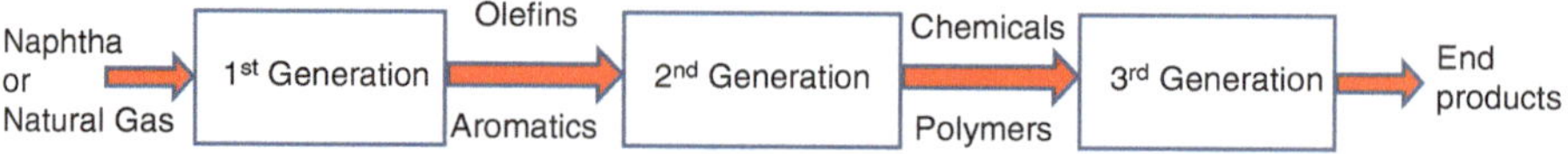

Fig. 4.1 Schematic representation of the petrochemical industry

Fig. 4.2 Industrial plant for the production of ethylene from ethanol in Triunfo, Brazil (reprinted with permission of Braskem)

which usually comprises C5–C10 aliphatic hydrocarbons. However, ethylene and propylene can also be produced from natural gas or even coal.

The second generation of the petrochemical industry involves the transformation of the olefins and aromatics in other products and polymers. There are many different processes and chemicals produced in the second generation of the petrochemical industry, such as phenol, styrene, acrylic acid and polypropylene, among others.

The third generation of the petrochemical industry involves the manufacture of end products, which go directly to the market or to the consumer. The major part of the third-generation industries is based on the physical transformation of polymers to produce goods and products to the consumer, not being straightforward considered a branch of the chemical industry.

Today, biomass can be used to produce different commodity chemicals. For instance, ethylene can be obtained from the dehydration of ethanol produced from sugar cane, corn or other carbohydrate sources. In Brazil, there is an industrial plant from Braskem (the major Brazilian petrochemical company) that runs a catalytic dehydration of ethanol over alumina-based catalyst (Fig. 4.2). The ethylene produced in this process is further transformed into polyethylene and commercialized

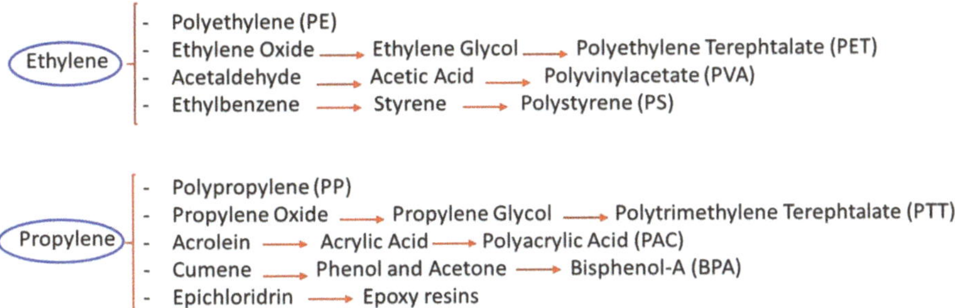

Fig. 4.3 Simplified tree of products obtained from ethylene and propylene

as a drop-in polymer. This means that the polyethylene obtained from ethanol has, essentially, the same properties and purity of the polymer obtained from naphtha or natural gas. Thus, biomass can be successfully used as raw material for the production of commodity chemicals.

The petrochemical tree of products has two major branches: those relying on ethylene and others based on propylene (Fig. 4.3). Bioethanol can afford the main products of the ethylene branch, whereas glycerol can fulfil the products of the propylene family, making what can be called the green petrochemistry. In the next subchapters, some developments of glycerol to commodity chemicals will be discussed.

4.2 Glycerol Hydrogenolysis

Hydrogenolysis can be defined as a splitting reaction to break C–C, C–O and other bonds, by the action of hydrogen (H_2). It usually requires metal catalysts, as the H–H bond is strong and difficult to be thermally broken. The use of a catalyst facilitates the breaking of the hydrogen molecule forming adsorbed hydrogen atoms on the surface, which can react further with the substrate.

Glycerol hydrogenolysis can afford products of C–O and C–C splitting yielding 1,2 and 1,3-propanediol, isopropanol, n-propanol, ethylene glycol, propylene and propane, among other products.

4.2.1 Glycerol to Propanediols

1,2-Propanediol, also known as propylene glycol, is an important chemical used as automotive antifreeze additive, de-icing fluid (Fig. 4.4), humectant for food, solvent for pharmaceutical applications and in the production of polymers, especially polyester resins. The world's production of propylene glycol was around 2.2 million tons in 2013, with most of the production made in the USA, China and Germany.

The traditional route is based on the hydrolysis of propylene oxide, which in turn is obtained from propylene (Scheme 4.1).

Fig. 4.4 De-icing of airplanes with propylene glycol-based fluid (Source: https://br.pinterest.com/pin/488007309598045787/)

Scheme 4.1 Petrochemical route of propylene glycol production

Propylene → Propylene Oxide → Propylene Glycol

The world's production of 1,3-propanediol is around 400,000 tons per year. Today, the main route of 1,3-propanediol production involves the fermentation of carbohydrates by a genetically modified strain of *E. coli*; this subject has been discussed in Chap. 3. Other traditional routes involve the hydration of acrolein and hydroformylation of ethylene oxide, followed by hydrogenation. The main use of 1,3-propanediol is in the reaction with terephthalic acid to yield polytrimethylene

Scheme 4.2 Hydrogenolysis of glycerol to propanediols

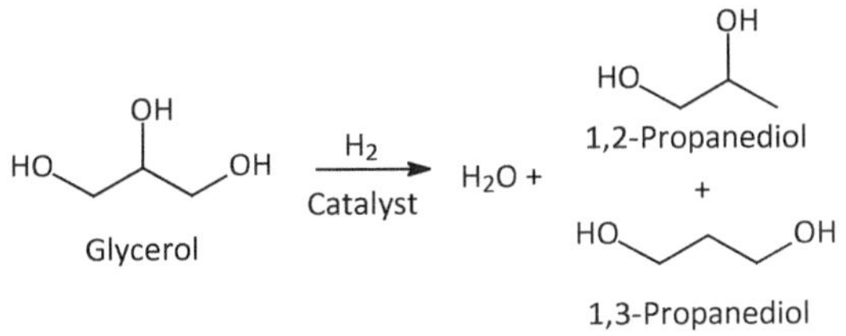

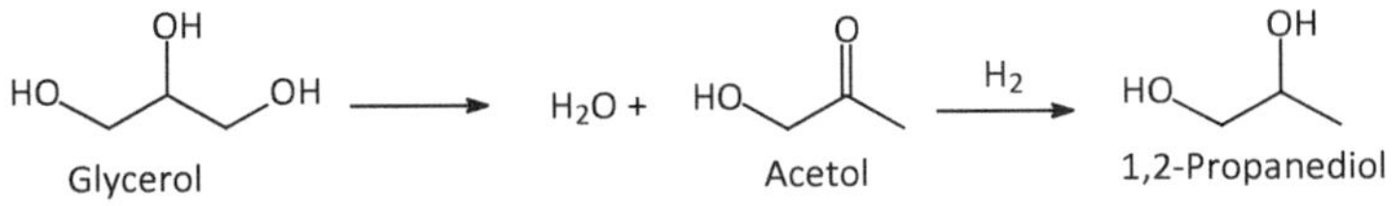

Scheme 4.3 Possible mechanistic pathway of glycerol hydrogenolysis to 1,2-propanediol

terephthalate (PTT), a polyester used in the production of carpets and textiles. The fibre is commercialized with the brand name of Corterra. It shows superior stain resistance, durability, softness and wide range of colours, because it holds dyes better than other conventional fibres.

The hydrogenolysis of glycerol may afford 1,2- and 1,3-propanediols (Scheme 4.2). Water is the other product of glycerol hydrogenolysis, but other oxygenated products can be formed, depending on the reaction conditions and catalysts.

The reaction can be carried out in batch or continuous flow conditions, with temperatures normally ranging from 180 to 250 °C. Metal-supported catalysts have been used [1], but copper chromite also leads to high conversion and selectivity to the desired product [2]. The concomitant use of an acidic catalyst, such as sulfonic acid resins, improves the selectivity towards hydrogenolysis, enabling the reaction to be carried out at lower temperatures and reduced pressures in batch conditions [3].

The 1,2-propanediol is usually produced in higher selectivity than the 1,3-isomer, but changing operational conditions, as well as the catalyst, can lead to 1,3 to 1,2 ratios of up to 5 [4]. The most acceptable mechanistic pathway involves dehydration of the glycerol molecule to hydroxy acetone (acetol), followed by hydrogenation to afford 1,2-propanediol (Scheme 4.3). Nevertheless, other mechanistic schemes have been proposed [5]. More severe reaction conditions (higher temperatures and pressure) lead to deeper hydrogenolysis of the glycerol molecule, yielding n-propanol and isopropanol [6, 7]. The reaction pathway is complex and may involve dehydration steps. CO and CO_2 can also be formed over supported Pt catalyst [8].

The hydrogenolysis of glycerol to 1,2-propanediol has been industrially implemented by ADM, which houses a 100,000 m^3 plant in the USA using proprietary technology. The Dow Chemical Company has developed a process named propylene glycol renewable (PGR), but to date, it has not been industrially implemented.

Table 4.1 shows some selected results of glycerol hydrogenolysis to 1,2- and 1,3-propanediols.

Table 4.1 Results of glycerol hydrogenolysis to 1,2- and 1,3-propanediols

Catalyst	Reaction conditions	Results (%)	Reference
Rh/C (5%)	20% glycerol solution, 180 °C, 80 bar, 168 h, 0.5 mmol Rh	[a]C = 21 [b]$S_{1,3\ PDO}$ = 12 [c]$S_{1,2\ PDO}$ = 70	[1]
Rh/Al_2O_3	20% glycerol solution, 180 °C, 80 bar, 168 h, 0.5 mmol Rh	C = 21 $S_{1,3\ PDO}$ = 12 $S_{1,2\ PDO}$ = 70	[1]
5% Ru/C	80% glycerol solution, 200 °C, 200 psi, 24 h	C = 44 $S_{1,2\ PDO}$ = 40	[2]
5% Ru/Al_2O_3	80% glycerol solution, 200 °C, 200 psi, 24 h	C = 23 $S_{1,2\ PDO}$ = 60	[2]
5% Pd/C	80% glycerol solution, 200 °C, 200 psi, 24 h	C = 29 $S_{1,2\ PDO}$ = 72	[2]
5% Pt/C	80% glycerol solution, 200 °C, 200 psi, 24 h	C = 35 $S_{1,2\ PDO}$ = 83	[2]
Ni/C	80% glycerol solution, 200 °C, 200 psi, 24 h	C = 40 $S_{1,2\ PDO}$ = 69	[2]
Ru/C	20% glycerol aqueous solution, 120 °C, 8 MPa, 10 h, 150 mg metal catalyst	C = 30 $S_{1,3\ PDO}$ = 5 $S_{1,2\ PDO}$ = 26	[3]
Rh/C	20% glycerol aqueous solution, 120 °C, 8 MPa, 10 h, 150 mg metal catalyst	C = 20 $S_{1,3\ PDO}$ = 7 $S_{1,2\ PDO}$ = 63 $S_{1\text{-}PO}$ = 19	[3]
Ru/C + Amberlyst	20% glycerol aqueous solution, 120 °C, 8 MPa, 10 h, 150 mg metal catalyst +300 mg Amberlyst	C = 31 $S_{1,3\ PDO}$ = 5 $S_{1,2\ PDO}$ = 55	[3]
$Pt/WO_3/ZrO_2$	10% glycerol solution in ethanol, 170 °C, 5.5 MPa, 12 h	C = 38 $S_{1,3\ PDO}$ = 23 $S_{1,2\ PDO}$ = 14	[4]
$Pt/WO_3/ZrO_2$	10% glycerol solution in water, 170 °C, 5.5 MPa, 12 h	C = 25 $S_{1,3\ PDO}$ = 26 $S_{1,2\ PDO}$ = 15	[4]
$Pt/WO_3/ZrO_2$	10% glycerol solution in ethanol/water, 170 °C, 5.5 MPa, 12 h	C = 46 $S_{1,3\ PDO}$ = 21 $S_{1,2\ PDO}$ = 8	[4]
Pt/C	1% glycerol solution, 200 °C, 40 bar, 5 h	C = 13	[9]
Pt/C	1% glycerol solution in 0.8 M CaO, 200 °C, 40 bar, 5 h	C = 100	[9]
Cu-ZnO	20% glycerol solution, 200 °C, 4.2 MPa, 12 h	C = 34 $S_{1,2\ PDO}$ = 78	[10]
Ru/C	20% glycerol solution, 180 °C, 8 MPa, 10 h, 15 mg Ru/C	C = 2 $S_{1,2\ PDO}$ = 43	[11]
Ru/C + Amberlyst 15	20% glycerol solution, 180 °C, 8 MPa, 10 h, 15 mg Ru/C, 140 mol H+ in the resin	C = 10 $S_{1,2\ PDO}$ = 44	[11]

(continued)

Table 4.1 (continued)

Catalyst	Reaction conditions	Results (%)	Reference
Ru/C + Amberlyst 70	20% glycerol solution, 180 °C, 8 MPa, 10 h, 15 mg Ru/C, 140 mol H^+ in the resin	C = 46 $S_{1,2\ PDO}$ = 48	[11]
Ru/C + BEA	20% glycerol solution, 180 °C, 8 MPa, 10 h, 150 mg Ru/C + 300 mg solid acid	C = 10 $S_{1,3\ PDO}$ = 3 $S_{1,2\ PDO}$ = 51	[12]
Ru/C + USY	20% glycerol solution, 180 °C, 8 MPa, 10 h, 150 mg Ru/C + 300 mg solid acid	C = 7 $S_{1,3\ PDO}$ = 0.4 $S_{1,2\ PDO}$ = 82	[12]
Ru/C + Amberlyst	20% glycerol solution, 180 °C, 8 MPa, 10 h, 150 mg Ru/C + 300 mg solid acid	C = 15 $S_{1,3\ PDO}$ = 2 $S_{1,2\ PDO}$ = 53	[12]
Ru/C + Amberlyst	20% glycerol solution, 140 °C, 8 MPa, 10 h, 150 mg Ru/C + 300 mg solid acid	C = 41 $S_{1,3\ PDO}$ = 1 $S_{1,2\ PDO}$ = 42	[12]

[a]C = Glycerol conversion
[b]$S_{1,3\ PDO}$ = Selectivity to 1,3-propanediol
[c]$S_{1,2\ PDO}$ = Selectivity to 1,2-propanediol

4.2.2 Glycerol to Propene

Propene or propylene is one of the main commodity chemicals, with worldwide production of over 90 million tons per year. The demand is expected to grow to 130 million tons by 2023. Today, most of the propene comes from fossil sources. Steam cracking of naphtha is the major process [13], but the olefin can also be produced from natural gas or coal through the methanol to olefin (MTO) [14] or methanol to propene (MTP) [15] processes. Another route involves the dehydrogenation of propane [16], which can be separated from the liquefied petroleum gas (LPG) fraction. About two thirds of the propene produced is directed to the manufacture of polypropylene, a thermoplastic resin of many uses.

The hydrotreatment of vegetable oils affords diesel-range hydrocarbons, CO, CO_2, methane, water and propane as major products. The reaction is conceptually a hydrogenolysis of the triglyceride molecule [17], ultimately leading to completely deoxygenated organic compounds, which can be blended to the traditional diesel oil. The propane molecule may be formed by the complete hydrogenolysis of the glycerol molecule and could be further dehydrogenated to afford propene. Nevertheless, this route would not be economically feasible, and a direct route of glycerol to propene must be pursued.

The hydrogenolysis of glycerol over Fe-Mo supported over activated carbon has been shown to give up to 90% selectivity to propene (Scheme 4.4) at proper reaction conditions [18–20]. The reaction can be carried out at the gas phase, under continuous flow conditions and atmospheric pressure, yielding 100% glycerol conversion. Other products, such as propane, arising from complete hydrogenolysis, ethane and propane, from C–C splitting, may also be observed.

Scheme 4.4 Hydrogenolysis of glycerol to propene

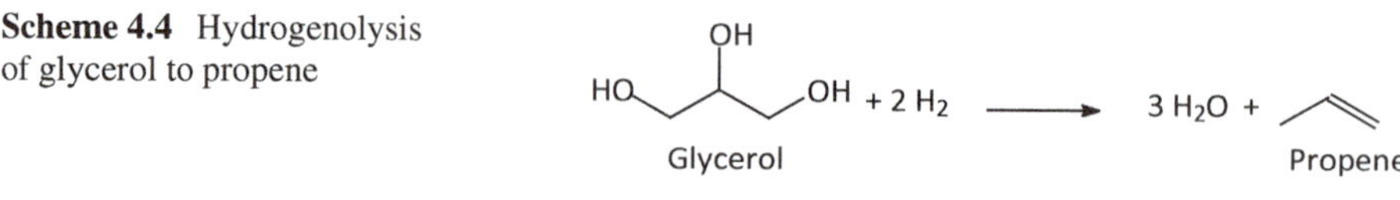

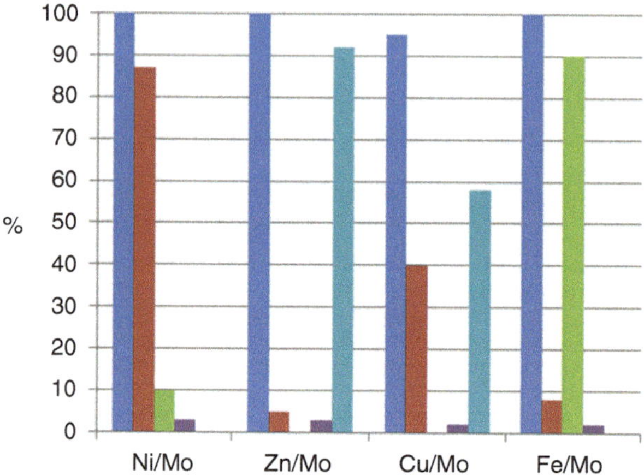

Fig. 4.5 Glycerol hydrogenolysis over metal/molybdenum supported on activated carbon at 300 °C, 1 atm, 5.4 h^{-1} WHSV and 120:1 H_2:glycerol molar ratio; (*blue-filled square*) conversion; (*red-filled square*) propane; (*green-filled square*) propene; (*purple-filled square*) methane and ethane; (*light blue-fille square*) oxygenated compounds. (Reprinted from Journal of Molecular Catalysis A, 422, Claudio J. A. Mota, Valter. L. C. Gonçalves, Jhon E. Mellizo, Ana Maria Rocco, José C. Fadigas, Rossano Gambetta. "Green Propene through the Selective Hydrogenolysis of Glycerol over Supported Iron-Molybdenum Catalyst: The Original History", 158–164., Copyright (2016), with permission from Elsevier)

Other molybdenum-based catalysts can also convert the glycerol molecule, but the selectivity to propene is low, or this olefin is not even observed (Fig. 4.5). The high selectivity of the Fe-Mo catalyst towards propene may reside on its low reducibility. X-ray diffraction (XRD) analysis showed the presence of metal oxide phases, including mixed Fe-Mo-O_x (Fig. 4.6), whereas temperature-programmed reduction (TPR) indicated significant reduction of the catalyst above 550 °C, which was the temperature used on the in situ reduction of the catalyst prior to glycerol hydrogenolysis. Hence, the final step of C = C hydrogenation was significantly suppressed, preventing the transformation of the formed propene into propane. At higher pressures and temperatures, however, such a reaction becomes favourable and the selectivity to propene decreases.

The active phase seems to be MoO_2 [21], and the mechanistic pathway may involve a bifunctional catalytic system, with dehydration and hydrogenation steps occurring on acidic and metallic sites, respectively. Acetol, 1,2-propanediol, acetone and isopropanol appear as potential intermediates (Scheme 4.5); these molecules can be observed at less severe reaction conditions, supporting the proposed mechanistic pathway from glycerol to propene.

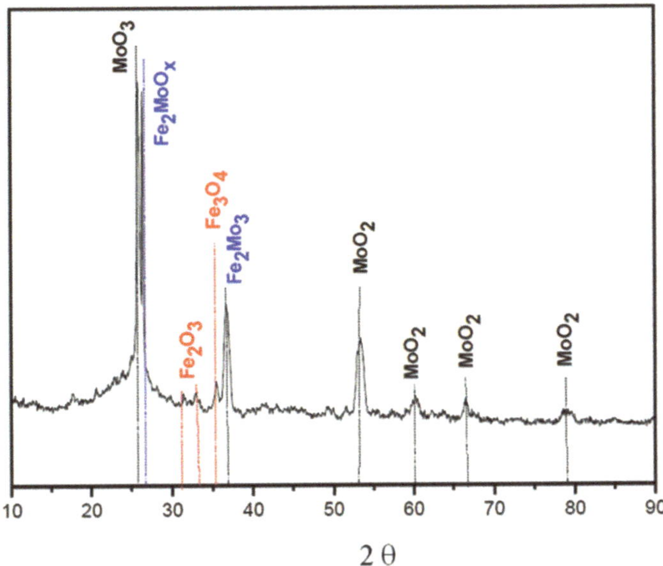

Fig. 4.6 XRD pattern of the Fe-Mo supported on activated carbon (Reprinted from Journal of Molecular Catalysis A, 422, Claudio J. A. Mota, Valter. L. C. Gonçalves, Jhon E. Mellizo, Ana Maria Rocco, José C. Fadigas, Rossano Gambetta. "Green Propene through the Selective Hydrogenolysis of Glycerol over Supported Iron-Molybdenum Catalyst: The Original History", 158–164., Copyright (2016), with permission from Elsevier)

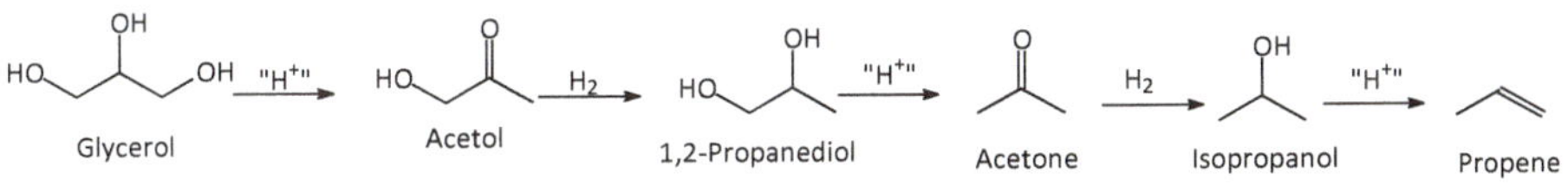

Glycerol —"H^+"→ Acetol —H_2→ 1,2-Propanediol —"H^+"→ Acetone —H_2→ Isopropanol —"H^+"→ Propene

Scheme 4.5 Possible mechanistic pathway of glycerol to propene, involving acid-catalysed dehydration and metal-catalysed hydrogenation steps

Propene can also be obtained from glycerol in two steps, using an iridium-supported catalyst and HZSM-5 zeolite [22]. The metal catalyst selectively produces isopropanol, whereas the zeolite catalyses the dehydration of the alcohol to propene.

4.3 Glycerol Dehydration

Dehydration of the glycerol molecule may follow two routes: dehydration of the primary hydroxyl leads to hydroxy acetone (acetol), whereas dehydration of the secondary hydroxyl gives 3-hydroxypropanal. This later molecule may undergo a second dehydration to produce acrolein (Scheme 4.6), which is a chemical of high industrial value, used in the production of acrylic acid and the amino acid

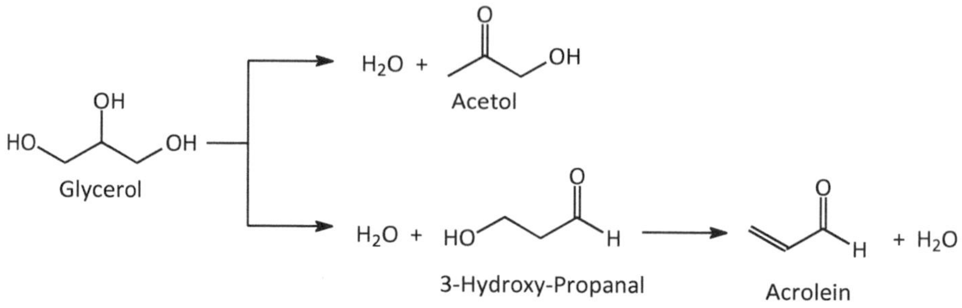

Scheme 4.6 Glycerol dehydration pathways leading to acetol and acrolein

methionine. Acrolein is also used as biocide in water treatment, especially to control algae. It is industrially produced from the oxidation of propene over bismuth- and molybdenum-based catalysts. Acetol finds applications in textile dye formulations, leather tanning, printing ink formulations and food flavouring, as well as intermediate in the synthesis of pharmaceuticals and cosmetics [23].

Glycerol can be dehydrated to acetol and glycerol over acid catalysts. Zeolites, metal oxides, clays and heteropoly acids have been tested, but conversion and selectivity depend on the catalysts and conditions used [24, 25]. In general, the best reported conversions range from 80 to 100%, with acrolein selectivity within 80–90%. Deactivation of the catalyst is a major problem, especially on highly acidic materials, such as zeolites. Acetol is usually formed as by-product of glycerol dehydration to acrolein, but it is the main product of glycerol dehydration when the reaction is performed at lower temperatures (260 °C), using La_2CuO_4 and copper-exchanged hydroxyapatite as basic catalysts [26, 27]. It has been suggested [28] that Brønsted acid sites favour the dehydration of the secondary hydroxyl of glycerol, leading to 3-hydroxypropanal and lately to acrolein. Lewis acidity directs the reaction to acetol, favouring dehydration of the primary hydroxyl of the glycerol molecule (Scheme 4.7).

Production of acrylic acid reached over 5.8 million tons in 2014, with sales of US$15.5 billion. The increasing demand for this chemical is being pushed by the personal care sector, where it is used in the form of superabsorbent polymer (sodium polyacrylate—Fig. 4.7), which is able to absorb about 200 times its weight in water. The use of acrylates in surfactants, adhesives, surface coatings and inks is also growing. A rapid growth in the acrylic acid market is expected until 2020, when the production will reach 8.5 million tons, with expected sales of US$22.5 billion. Thus, the search for renewable routes of acrylic acid production is highly desirable.

Acrylic acid is industrially produced from the oxidation of acrolein over molybdenum and vanadium oxide catalysts [29, 30], through the Mars-van Krevelen mechanism [31], in which an oxygen atom from the structure of the catalyst is added to the substrate, leaving a vacancy that is further completed by molecular oxygen from the reaction medium.

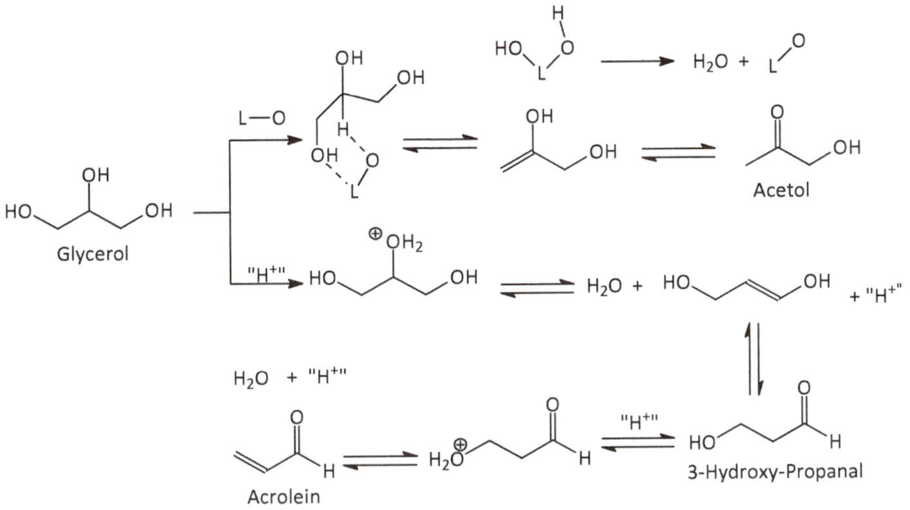

Glycerol

Acetol

Acrolein

3-Hydroxy-Propanal

Scheme 4.7 Proposed role of Brønsted and Lewis acid sites in the dehydration of the secondary and primary hydroxyls of glycerol (adapted from Ref. [28])

Fig. 4.7 Sodium polyacrylate—water balls (Source: https://br.pinterest.com/pin/529454499924361479/)

The dehydration of glycerol to acrolein can be used to develop a process for the production of renewable acrylic acid, from biomass material. Nevertheless, the main economic feasibility of such a process would come from the direct transformation of glycerol to acrylic acid over bifunctional catalysts, having acidic and redox sites. The task is difficult, because dehydration is endothermic, whereas oxidation is exothermic. Thus, controlling the reaction conditions to maximize both transformations is not simple.

There are some studies in the literature, covering mixed inorganic oxides, metal-impregnated zeolites and heteropoly acids [32–37] directed to the glycerol oxidehydration in the presence of oxygen or air (Scheme 4.8). Although the glycerol

Scheme 4.8 Glycerol oxidehydration to acrylic acid over bifunctional catalysts, having acidic and redox sites

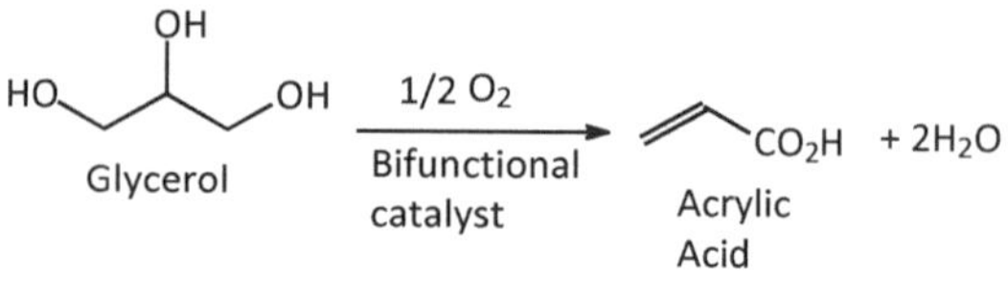

conversion is usually above 70%, the selectivity to acrylic acid ranges from 5 to 35%, suggesting that the redox function of the catalysts is still not well adjusted or the reaction conditions do not favour oxidation. Acrolein is reported to be the main product of glycerol reaction in most of the catalyst used for oxidehydration.

4.4 Glycerol Halogenation to Epichlorohydrin

The world's production of epichlorohydrin accounts for approximately 2 million tons per year. This chemical is mostly converted to epoxy resins for the use in paints, coating and adhesives (Fig. 4.8). Other uses include the synthesis of glycidyl nitrate (a binder for explosives and propellant), as inset fumigant and in the production of ion-exchange resins, plasticizers, dyestuffs and pharmaceutical products, as well as in water treatment and the paper industry.

Traditionally, epichlorohydrin has been produced in three steps from propene. Initially, the olefin reacts with chlorine at high temperatures, undergoing free-radical chemistry to produce allyl chloride, which is then treated with hypochlorous acid (HOCl) to yield a mixture of 1,2-dichloro-1-propanol (70%) and 1,3-dichloro-2-propanol (30%) [38]. Treatment of the halohydrin products with sodium hydroxide gives epichlorohydrin (Scheme 4.9). A drawback of the process is that the 1,2-dichloro isomer is tenfold less reactive towards the base to yield epichlorohydrin than the 1,3-dichloro analogue.

Solvay S.A., a French-Belgian chemical company, has developed a process named Epicerol to convert glycerol into epichlorohydrin. The company claims that this process reduces in 61% the global greenhouse gas emission relative to the oil-based process. This account includes the use of renewable carbon source (glycerol) and less energy consumption. Each ton of epichlorohydrin produced in the Epicerol process reduces the carbon footprint in 2.56 tons of CO_2 equivalent. In addition, less-chlorinated wastes are discharged, together with lower inputs of water and chlorinated compounds.

The process involves the reaction of glycerol with hydrochloric acid (HCl), in the presence of Brønsted or Lewis acid catalysts, to afford almost exclusively the 1,3-dichloro-2-propanol [39], which is then treated with sodium hydroxide to yield epichlorohydrin (Scheme 4.10). The first chlorination yields 1-chloro-1,2-propanediol (chlorination of the primary hydroxyl of glycerol) as the main product, together with minor amount of 2-chloro-1,3-propanediol (chlorination of the secondary hydroxyl of glycerol). Nevertheless, the second chlorination is strongly

Fig. 4.8 Epichlorohydrin-based epoxy resins (Source: http://losabalorios.com/blog/2014/07/como-reparar-recipientes-de-arcilla/)

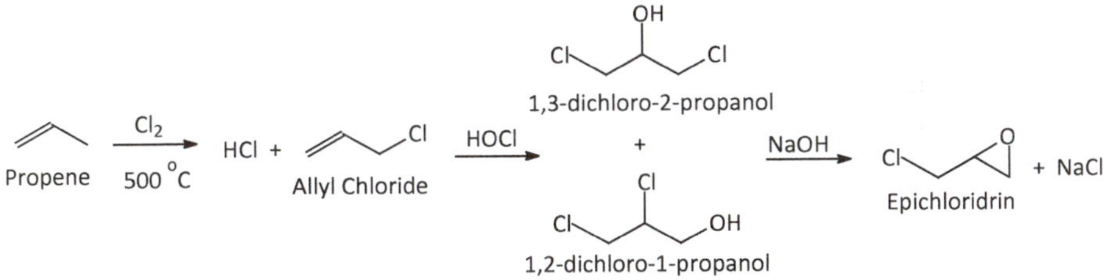

Propene $\xrightarrow[500\,^{\circ}C]{Cl_2}$ HCl + Allyl Chloride $\xrightarrow{HOCl}$ 1,3-dichloro-2-propanol + 1,2-dichloro-1-propanol $\xrightarrow{NaOH}$ Epichloridrin + NaCl

Scheme 4.9 Traditional industrial synthesis of epichlorohydrin from propene

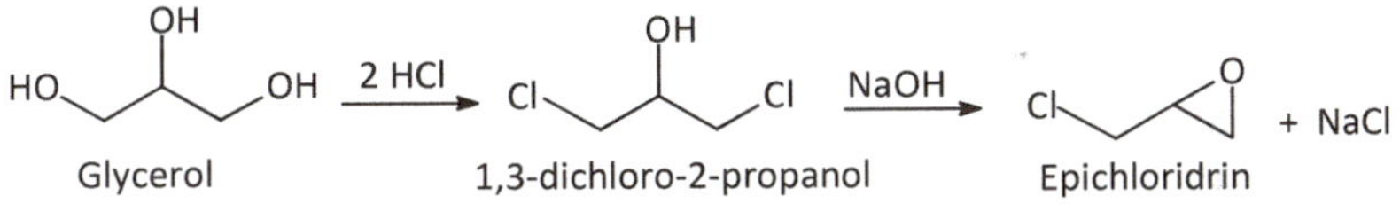

Glycerol $\xrightarrow{2\ HCl}$ 1,3-dichloro-2-propanol $\xrightarrow{NaOH}$ Epichloridrin + NaCl

Scheme 4.10 Production of epichlorohydrin from glycerol

inhibited in this later isomer, and the 1,3-dichloro-2-propanol is obtained in high selectivity from the 1-chloro-1,2-propanediol [39]. This different behaviour brings some advantages to the Epicerol process compared with the traditional route from propene, as the reactivity of the 1,3-dichloro halohydrin is highly superior than the reactivity of the 1,2-dichloro isomer.

Solvay S.A. has plants running the Epicerol process in Thailand and China, both producing 100,000 tons of epichlorohydrin per year. A recent techno-economic analysis of the process has been reported [40]. The process simulation was carried out to run in a semi-batch model, producing 26,500 tons per year of epichlorohydrin. At this scale, the total investment costs were around 64 million euros with a production cost of epichlorohydrin of 1.28 euros per kg. The pay-out time was estimated in 5 years, whereas the plant has been projected to run during 20 years.

4.5 Glycerol Reforming

Reforming is a general process that involves the reaction of an organic compound with water vapour, usually at high temperatures, to mostly yield CO and H_2, known as synthesis gas or syngas. Coal reforming has been known since the nineteenth century as a technological pathway to produce hydrogen. However, in the twentieth century, natural gas reforming has become the major route to syngas, which is used in the synthesis of ammonia, methanol, hydroformylations and the Fischer-Tropsch synthesis to produce hydrocarbons. When hydrogen is the main target, the shift process is also necessary, because CO is a poison to many hydrogenation catalysts. Syngas reacts with water over copper- and chromium-based catalysts to convert CO into CO_2, producing additional hydrogen (Scheme 4.11). Natural gas reforming together with the shift process accounts for nearly 80% of the H_2 produced in the world. Today, due to the importance of carbon capture, sequestration and utilization (CCSU), this route is named pre-combustion and is being industrially implemented to facilitate the CO_2 capture, allowing the hydrogen gas to be burned or used in fuel cell applications [41].

The glycerol reforming produces mainly hydrogen, carbon dioxide and carbon monoxide. The reaction is usually carried out between 500 and 750 °C in the presence of metal-supported catalysts (Ni, Pt, Ru, Co) and water-to-glycerol molar ratio between 6 and 9 [42]. Scheme 4.12 shows the thermodynamic of the reactions.

Aqueous-phase glycerol reforming can be carried out at 350 °C and 60 bar over Pt supported on CeO_2/ZrO_2 catalyst [43]. Supported Ni [44] and mixed metal-supported systems [45] can also be employed in the aqueous-phase reforming of glycerol. The product is syngas, a mixture of CO and H_2. Coupling of aqueous-phase reforming with Fischer-Tropsch synthesis yields diesel-range hydrocarbons [46].

Hydrogen can also be produced by the supercritical water reforming. The main advantage is the higher reactivity of water at the supercritical point. Yet, CO_2 is

Scheme 4.11 Production of hydrogen from natural gas

$$CH_4 + H_2O \longrightarrow CO + 3H_2 \quad \Delta H = +192\ kJ/mol \quad \text{(Reforming)}$$

$$CO + H_2O \longrightarrow CO_2 + H_2 \quad \Delta H = -41\ kJ/mol \quad \text{(Shift)}$$

$$CH_4 + 2H_2O \longrightarrow CO_2 + 4H_2 \quad \Delta H = +151\ kJ/mol$$

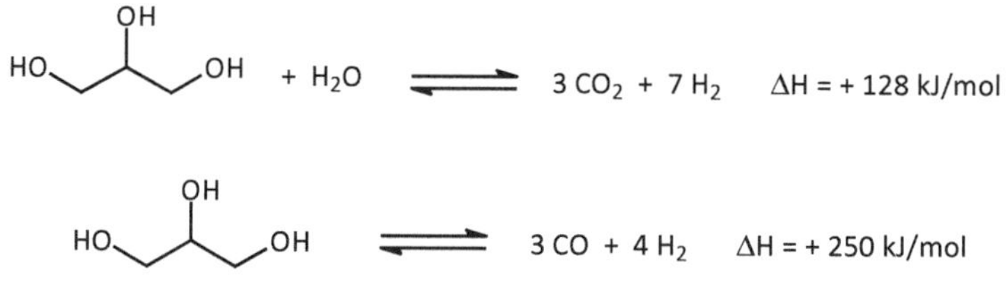

$$\text{HOCH}_2\text{CH(OH)CH}_2\text{OH} + H_2O \rightleftharpoons 3CO_2 + 7H_2 \quad \Delta H = +128\ kJ/mol$$

$$\text{HOCH}_2\text{CH(OH)CH}_2\text{OH} \rightleftharpoons 3CO + 4H_2 \quad \Delta H = +250\ kJ/mol$$

Scheme 4.12 Thermodynamics of glycerol reforming

preferentially formed over CO. On the other hand, temperature of 600 °C and above is necessary to achieve high H_2 production, making the process less attractive.

4.6 Glycerol to Methanol

Because biodiesel is normally obtained from the transesterification of triglycerides with methanol, a process to convert glycerol in methanol would be highly desirable. This could be a good way to make the biodiesel process more sustainable, without using any source of fossil-derived raw material.

The traditional process of methanol synthesis involves natural gas reforming to syngas, followed by CO hydrogenation over Cu-ZnO-based catalysts [47] (Scheme 4.13). More recently, hydrogenation of CO_2 has been industrially applied in the production of methanol [48]. Carbon Recycling International uses geothermic energy to generate the hydrogen necessary to react with CO_2 to yield methanol, having an industrial plant in Iceland that uses this technology.

A company named BioMCN has developed a demonstration plant in the Netherlands to produce methanol from glycerol reforming [49]. The company adapted a traditional natural gas steam-reforming unit to operate with glycerol. Nevertheless, the production of methanol from glycerol has been discontinued in 2013, due to technical and economic reasons. The company still produces methanol, but using biogas as the major renewable source.

The process consists of a reformer sector and a methanol synthesis sector [50] (Fig. 4.9). Glycerol and water are injected in the reactor, maintained around 700 °C and 240–270 bar. Water is condensed at the reactor outlet and the gas phase directed to the methanol synthesis reactor, kept at 250 °C and at the same range of pressure of the steam-reforming reactor. The company was able to transform up to 60% of the carbon atoms of the glycerol molecule in methanol, which is close to the thermodynamic equilibrium at the conditions used.

Hydrogenolysis of glycerol over molybdenum or tungsten supported catalysts at 250–325 °C, and 60 bar of pressure affords different alcohols as products. Methanol was the main product over supported tungsten catalysts, whereas 1,2-propanediol was formed as the major product on molybdenum [51].

The reaction of aqueous glycerol over MgO or CaO at temperatures between 280 and 340 °C yields methanol as the major product. Ethanol, ethanal, propanal and 2,3-butanedione were also formed [52]. The proposed mechanism (Scheme 4.14)

$$CH_4 + H_2O \longrightarrow CO + 3\,H_2 \qquad \Delta H = +49\ \text{kcal/mol}$$

$$CO + 2\,H_2 \rightleftharpoons CH_3OH \qquad \Delta H = -22\ \text{kcal/mol}$$

$$CO_2 + 3\,H_2 \rightleftharpoons CH_3OH + H_2O \qquad \Delta H = -12\ \text{kcal/mol}$$

Scheme 4.13 Traditional methanol synthesis processes, based on natural gas reforming or CO_2 hydrogenation

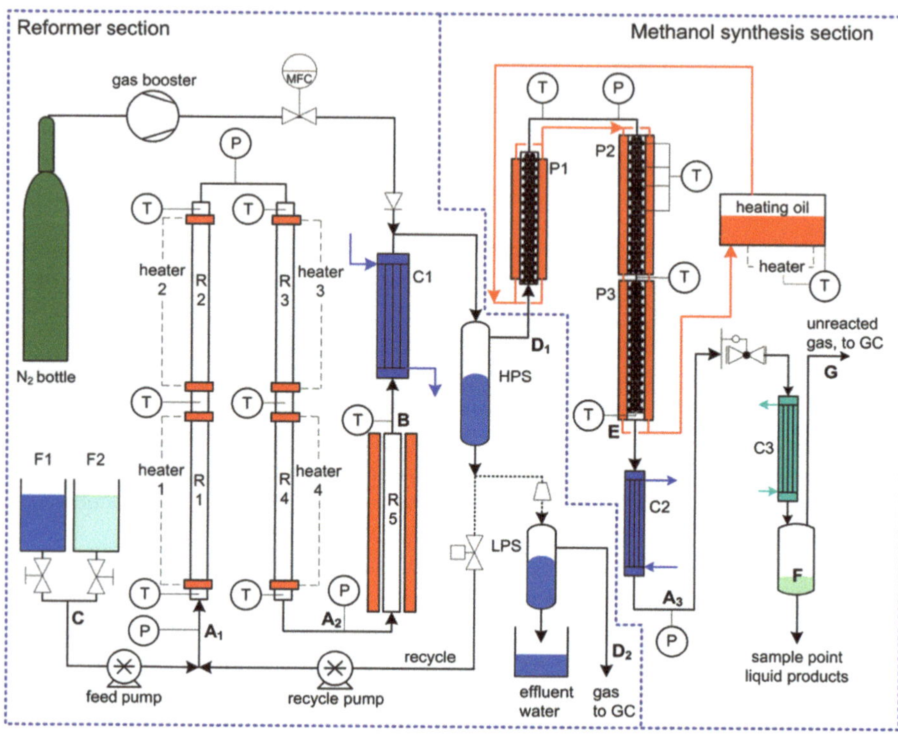

Fig. 4.9 Flow sheet of the integrated glycerol-to-methanol bench scale unit. HPS and LPS refer to high-pressure separator and low-pressure separator, respectively. *F* feed container, *C* cooler, *P* packed bed reactor, *R* reforming reactor. (© 2012 Van Bennekom JG, Venderbosch RH, Heeres HJ. Published in [Biomethanol from Glycerol. Biodiesel—Feedstocks, Production and Applications Zhen Fang (Ed), ISBN 978–953–51-0910-5, 498 pages, Publisher: InTech, Chapters published] under CC BY 3.0 licence. Available from: http://dx.doi.org/10.5772/53691)

Scheme 4.14 Possible mechanistic pathway of glycerol decomposition to methanol over alkaline-earth-metal oxide catalysts seems to involve the basic-catalysed decomposition of glycerol, yielding acetol, which can be further transformed into methanol and the acetyl radical. Another pathway involves the decomposition of glycerol to methanol and hydroxyacetaldehyde, which can be further decomposed into methanol and formaldehyde.

References

1. Chaminand J, Dajakovitch L, Gallezot P, Marion P, Pinel C, Rosierb C (2004) Glycerol hydrogenolysis on heterogeneous catalysts. Green Chem 6:359–361
2. Dasari MA, Kiatsimkul PP, Sutterlin WR, Suppes J (2005) Low-pressure hydrogenolysis of glycerol to propylene glycol. Appl Catal A 281:225–231
3. Miyazawa T, Kusunoki Y, Kunimori K, Tomishige K (2006) Glycerol conversion in the aqueous solution under hydrogen over Ru/C plus an ion-exchange resin and its reaction mechanism. J Catal 240:213–221
4. Gong L, Lu Y, Ding Y, Lin R, Li J, Dong W, Wang T, Chen W (2009) Solvent effect on selective dehydroxylation of glycerol to 1,3-propanediol over a Pt/WO3/ZrO2 catalyst. Chin J Catal 30:1189–1191
5. Zheng Y, Chen X, Shen Y (2008) Commodity chemicals derived from glycerol, an important biorefinery feedstock. Chem Rev 108:5223
6. Casale B, Gomez AM (1994) Catalytic method of hydrogenating glycerol. US pat 5, 276, 18, 1994
7. Ludwig S, Manfred E (1997) Preparation of 1,2-propanediol. US 5616817 A
8. Wawrzetz A, Peng B, Hrabar A, Jentys A, Lemonidou AA, Lercher JA (2010) Towards understanding the bifunctional hydrodeoxygenation and aqueous phase reforming of glycerol. J Catal 269:411–420
9. Maris EP, Davis RJ (2007) Hydrogenolysis of glycerol over carbon-supported Ru and Pt catalysts. J Catal 249:328–337
10. Wang S, Liu H (2007) Selective hydrogenolysis of glycerol to propylene glycol on cu–ZnO catalysts. Catal Lett 117:62–67
11. Miyazawa T, Koso S, Kunimori K, Tomishige K (2007) Glycerol hydrogenolysis to 1,2-propanediol catalyzed by a heat-resistant ion-exchange resin combined with Ru/C. Appl Catal A Gen 329:30–35
12. Kusunoki Y, Miyazawa T, Kunimori K, Tomishige K (2005) Highly active metal–acid bifunctional catalyst system for hydrogenolysis of glycerol under mild reaction conditions. Catal Commun 6:645–649
13. Ren T, Patel M, Blok K (2006) Olefins from conventional and heavy feedstocks: energy use in steam cracking and alternative processes. Energy 31:425
14. Chen D, Moljord K, Holmen A (2012) A methanol to olefins review: diffusion, coke formation and deactivation on SAPO type catalysts. Microporous Mesoporous Mater 164:239
15. Khanmohammadi M, Amani S, Bagheri Garmarudi A, Niaei A (2016) Methanol-to-propylene process: perspective of the most important catalysts and their behavior. Chin J Catal 37:325–339
16. Han Z, Li S, Jiang F, Wang T, Ma X, Gongo J (2014) Propane dehydrogenation over Pt–cu bimetallic catalysts: the nature of coke deposition and the role of copper. Nanoscale 6:10000–10008
17. Liu Y, Sotelo-Boyás R, Murata K, Minowa T, Sakanishi K (2011) Hydrotreatment of vegetable oils to produce bio-hydrogenated diesel and liquefied petroleum gas fuel over catalysts containing sulfided Ni–Mo and solid acids. Energy Fuel 25:4675–4685
18. Mota CJA, Gonçalves VLC, Fadigas JC, Gambetta R (2009) Preparation of heterogeneous catalysts used in selective hydrogenation of glycerin to propene, and a process for the selective hydrogenation of glycerin to propene. WO2009155674A1

19. Mota CJA, Gonçalves VLC, Fadigas JC, Gambetta R (2011) Preparation of heterogeneous catalysts used in selective hydrogenation of glycerin to propene, and a process for the selective hydrogenation of glycerin to propene. US20110184216A1
20. Mota CJA, Gonçalves VLC, Mellizo JE, Rocco AM, Fadigas JC, Gambetta R (2016) Green propene through the selective hydrogenolysis of glycerol over supported iron-molybdenum catalyst: the original history. J Mol Catal A 422:158–164
21. Zacharopoulou V, Vasiliadou ES, Lemonidou A (2015) One-step propylene formation from bio-glycerol over molybdena-based catalysts. Green Chem 17:903–912
22. Yu L, Yuan J, Zhang Q, Liu YM, He HY, Fan KN, Cao Y (2014) ChemSusChem 7:743
23. Mohamad MH, Awang R, Yunus WMZW (2011) A review of acetol: application and production. Am J Appl Sci 8:1135–1139
24. Chai SH, Wang HP, Liang Y, Xu BQ (2007) Sustainable production of acrolein: investigation of solid acid–base catalysts for gas-phase dehydration of glycerol. Green Chem 9:1130–1136
25. Katryniok B, Paul S, Bellière-Baca V, Rey P, Dumeignil F (2010) Glycerol dehydration to acrolein in the context of new uses of glycerol. Green Chem 12:2079–2098
26. Velasquez M, Santamaria A, Dupeyrat CB (2014) Selective conversion of glycerol to hydroxyacetone in gas phase over La_2CuO_4 catalyst. Appl Catal B 160-161:606–613
27. Oliveira AC, Carvalho DC, Pinheiro LG, Campos A, Miller ERC, Souza FF, Filho JM, Saraiva GD, Filho ACS, Fonseca MG (2014) Characterization and catalytic performances of copper and cobalt-exchanged hydroxyapatite in glycerol conversion for 1-hydroxyacetone production. Appl Catal A 14(471):39–49
28. Possato LF, Diniz RN, Garetto T, Pulcinelli SH, Santilli CV, Martins L (2013) A comparative study of glycerol dehydration catalyzed by micro/mesoporous MFI zeolites. J Catal 300:102–112
29. Andrushkevich TV (1993) Heterogeneous catalytic oxidation of acrolein to acrylic acid: mechanism and catalysts. Catal Rev Sci Eng 35:213–252
30. Kampe P, Giebelder L, Smuelis D, Kunert J, Drochner A, Haass F, Adams AH, Ott J, Endres S, Shimanke G, Buhrmester T, Martin M, Fuess H, Vogel H (2007) Heterogeneously catalysed partial oxidation of acrolein to acrylic acid—structure, function and dynamics of the V–Mo–W mixed oxides. Phys Chem Chem Phys 9:3577–3589
31. Doornkamp C, Ponec V (2000) The universal character of the Mars and van Krevelen mechanism. J Mol Catal A 162:19–32
32. Wang F, Dubois JL, Ueda W (2010) Catalytic performance of vanadium pyrophosphate oxides (VPO) in the oxidative dehydration of glycerol. Appl Catal A 376:25–321
33. Ulgen A, Hoelderich W (2011) Conversion of glycerol to acrolein in the presence of WO_3/TiO_2 catalysts. Appl Catal A 400:34–38
34. Soriano MD, Concepcion P, Nieto JML, Cavani F, Guidetti S, Trevisanut C (2011) Tungsten-vanadium mixed oxides for the oxidehydration of glycerol into acrylic acid. Green Chem 13:2954–2962
35. Pestana CFM, Guerra ACO, Ferreira GB, Turci CC, Mota CJA (2013) Oxidative dehydration of glycerol to acrylic acid over vanadium-impregnated zeolite beta. J Braz Chem Soc 24:100–105
36. Possato LG, Cassinelli WH, Garetto T, Pulcinelli SH, Santilli CV, Martins L (2015) One-step glycerol oxidehydration to acrylic acid on multifunctional zeolite catalysts. Appl Catal A Gen 492:243–251
37. Li X, Zhang Y (2016) Oxidative dehydration of glycerol to acrylic acid over vanadium-substituted cesium salts of keggin-type heteropolyacids. ACS Catal 6:2785–2791
38. Santacesaria E, Tesser R, Di Serio M, Casale L, Verde D (2010) New process for producing epichloridrin via glycerol chlorination. Ind Eng Chem Res 49:964–970
39. Tesser R, Santacesaria E, Di Serio M, Di Nuzzi G, Fiandra V (2007) Kinetics of glycerol chlorination with hydrochloric acid: a new route to α,γ-dichlorohydrin. Ind Eng Chem Res 46:6456–6465
40. Almena A, Martin M (2016) Technoeconomic analysis of the production of epichloridrin from glycerol. Ind Eng Chem Res 55:3226–3238

41. Leung DYC, Caramanna G, Mercedes Maroto-Vale M (2014) An overview of current status of carbon dioxide capture and storage technologies. Renew Sust Energ Rev 39:426–443
42. Schwengber CA, Alves HJ, Schaffner RA, Silva FA, Sequinel R, Bach VR, Ferracin RJ (2016) Overview of glycerol reforming for hydrogen production. Renew Sust Energ Rev 58:259–266
43. Soares RR, Simonetti DA, Dumesic JA (2006) Glycerol as a source for fuels and chemicals by low-temperature catalytic processing. Angew Chem Int Ed 45:3982
44. Shao S, Shi AW, Liu CL, Yang RZ, Dong WS (2014) Hydrogen production from steam reforming of glycerol over Ni/CeZrO catalysts. Fuel Process Technol 125:1–7
45. Kim SH, Go YJ, Park NC, Kim JH, Kim YC, Moon DJ (2015) Steam reforming of glycerol over nano size Ni-Ce/LaAlO3 catalysts. J Nanosci Nanotechnol 15(1):522–526
46. Simonetti DA, Rass-Hansen J, Kunkes EL, Soares RR, Dumesic JA (2007) Coupling of glycerol processing with Fischer–Tropsch synthesis for production of liquid fuels. Green Chem 9:1073
47. Olah GA, Goeppert A, Prakash GKS (2009) Beyond oil and gas: the methanol economy, 2nd edn. WileyVCH, Weinheim
48. Goeppert A, Czaun M, Jones JP, Prakash GKS, Olah GA (2014) Recycling of carbon dioxide to methanol and derived products – closing the loop. Chem Soc Rev 43:7995–8048
49. van Bennekom JG, Venderbosch RH, Assink D, Lemmens KPJ, Heere HJ (2012) Bench scale demonstration of the Supermethanol concept: the synthesis of methanol from glycerol derived syngas. Chem Eng J 207–208:245–253
50. van Bennekom JG, Venderbosch RH, Heeres HJ (2012) Biomethanol from glycerol. In: Fang Z (ed) Biodiesel – feedstocks, production and applications. InTech, Rijeka. 498 pp. ISBN: 978-953-51-0910-5
51. Shozi ML, Dasireddy VDBC, Singh S, Mohlala P, Morgan DJ, Friedrich HB (2016) Hydrogenolysis of glycerol to monoalcohols over supported Mo and W catalysts. ACS Sustain Chem Eng 4(10):5752–5760
52. Haider MH, Dummer NF, Knight DW, Jenkins RL, Howard M, Moulijn J, Taylor SH, Hutchings GJ (2015) Efficient green methanol synthesis from glycerol. Nat Chem 7:1028–1032

Chapter 5
Thermochemical Routes of Glycerol Transformation in Specialty Chemicals

Abstract Glycerol can be converted in ethers, acetals/ketals, esters and carbonate that find applications as fuel additives, special solvents and antioxidant, among others. The reaction of glycerol with ketone and aldehydes leads to the formation of cyclic oxygenated compounds, such as solketal, which is used as a solvent and plasticizer in the polymer industry or as suspension agent in pharmaceutical preparations. It can also be used as a fuel additive. Glycerol ethers comprehend compounds obtained in the reaction of glycerol with alkenes, such as isobutene, or alcohols, such as ethanol. Ether derivatives of glycerol are compatible with diesel and biodiesel fuels, being additives for these fuels. Glycerol dimer and trimer are used as starting materials for the production of emulsifiers in food and cosmetics. The glycerol acetates, monoacetin, diacetin and triacetin have many industrial applications, such as in cryogenics, synthesis of biodegradable polyesters and cosmetics. Glycerol carbonate is a relatively new product that is gaining new applications. Finally, oxidation of glycerol affords different products of high-added value.

Keywords Glycerol • Solketal • Acetal • Glycerol ethers • Fuel additive • Polyglycerol • Glycerol esters • Triacetin • Glycerol carbonate • Dihydroxyacetone

5.1 Specialty and Fine Chemicals

The fine chemical industry is characterized by the relatively low volume of production, in comparison with the traditional petrochemical or basic chemical industry. The products are usually used as dyes, pharmaceuticals, additives and many other special applications. Therefore, they have high-added value, usually higher than US$ 10.00 per kg and are produced in multipurpose batch units. The world market of fine chemicals was estimated in US$ 85 billion in 2010 [1].

Glycerol can be transformed in ethers, acetals/ketals and esters, all of them with many important applications. Glycerol carbonate is another product of this category, a relatively new substance that finds new applications every year. Last but not least, glycerol oxidation affords different products. Among them is dihydroxyacetone, which is the main constituent of self-tanning products.

C.J.A. Mota et al., *Glycerol*, DOI 10.1007/978-3-319-59375-3_5

5.2 Glycerol Ketal and Acetal

Glycerol ketals and acetals can be produced through the acid-catalysed reaction with ketones and aldehydes, respectively. The reaction of glycerol with acetone provides mainly one isomer, named (2,2-dimethyl-1,3-dioxane-4-yl) methanol, also known as solketal (Scheme 5.1). On the other hand, the reaction of glycerol with formaldehyde solution affords a mixture of two isomers named (1,3-dioxane-4-yl) methanol (five-membered ring acetal) and 1,3-dioxane-5-ol (six-membered ring acetal) (Scheme 5.2).

Acetalization/ketalization is a reversible reaction. Thus, excess of one of the reactants is usually employed, as well as the use of physical methods to remove the water formed. In fact, water may also interact with the catalyst, weakening the acidity, which may contribute to slow down the reaction rate.

5.2.1 Glycerol Acetals Through the Reaction with Aldehydes

The glycerol six-membered ring acetals are potential precursors for the production of the green platform chemicals, such as dihydroxyacetone and 1,3-propanediol that might be obtained by subsequent selective oxidation or hydrogenation, respectively. The condensation of glycerol with aldehydes yields a mixture of isomers with predominance of the six-membered cyclic acetal. This is due to the thermodynamic stability of the isomers. Although cyclization to form the five-membered ring isomer is kinetically favoured [2], their interchange does occur rapidly, and the thermodynamic distribution is usually observed, favouring the six-membered ring acetal.

The catalytic condensation of glycerol with benzaldehyde and formaldehyde was studied using Amberlyst-36, H-Beta and K-10 Montmorillonite as catalysts. The acetal isomers were produced in over 90% yield after 4 h [3].

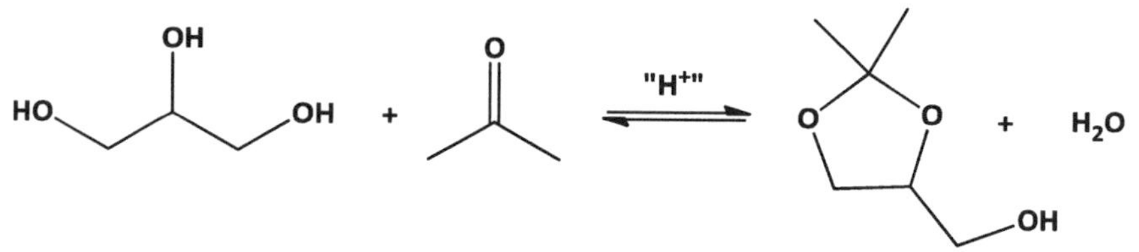

Scheme 5.1 Reaction of glycerol with acetone under acid catalysis conditions

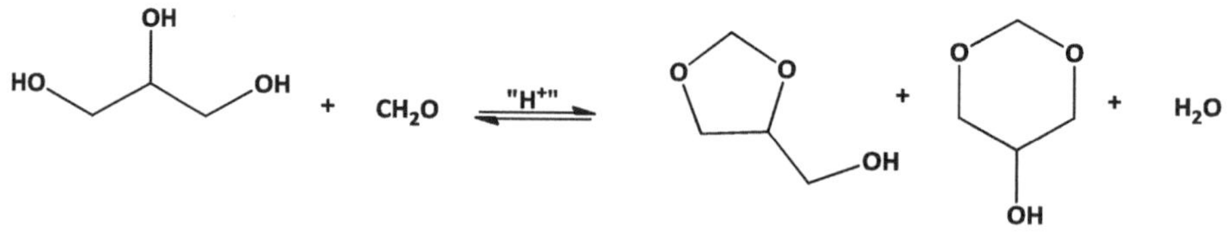

Scheme 5.2 Reaction of glycerol with formaldehyde under acid catalysis conditions

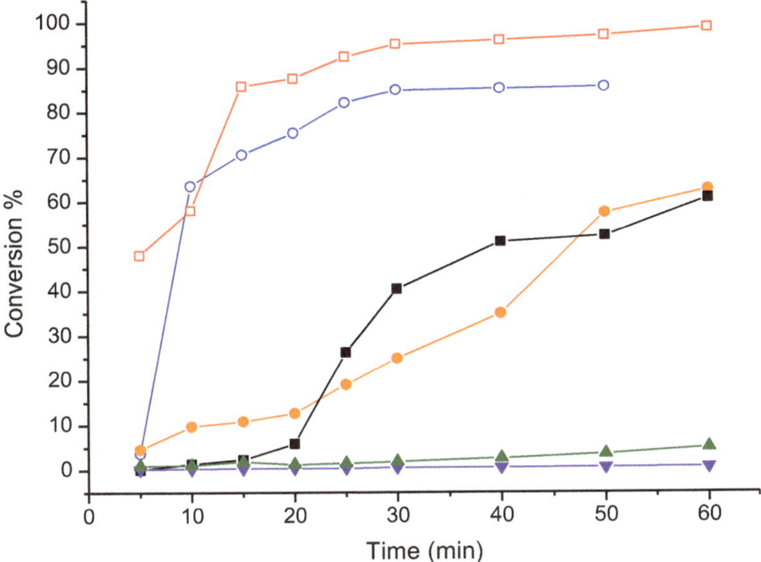

Fig. 5.1 Kinetics of the glycerol reaction with aqueous formaldehyde at 70°C over various acid catalysts. (*black-filled square*) Amberlyst-15, (*red open square*) zeolite Beta, (*orange-filled circle*) K-10 Montmorillonite, (*blue open circle*) PTSA, (*violet-filled inverted triangle*) HUSY, (*green-filled triangle*) zeolite ZSM-5 (Reprinted from Ref. [4] with the permission of the RSC)

The acid-catalysed reaction of glycerol with aqueous formaldehyde in the absence of solvents was studied using heterogeneous catalysts, such as Amberlyst-15, K-10 Montmorillonite and zeolites HUSY, HZSM-5 and H-Beta [4]. A homogeneous system, p-toluenesulfonic acid, was also used for comparison purpose. The glycerol conversion against time for each catalyst is shown in Fig. 5.1. Zeolites HUSY and HZSM-5 showed negligible conversion. This later zeolite presents shape selective properties [5], which may partly explain the results. HUSY is, however, a large-pore zeolite, and the low activity cannot be explained by the concept of shape selectivity. On the other hand, zeolite H-Beta showed the best performance among all catalysts tested, converting around 95% of the glycerol within 60 min of reaction time. Zeolite Beta is also a large-pore zeolite, but has a hydrophobic character, due to the high Si/Al ratio, around 16. This may explain the results, because the pore environment of zeolite Beta prevents the diffusion of the water from the solution to the interior of the pore, preserving the strength of the acid sites. Furthermore, the water formed during acetalization is expelled off from the pore environment, reducing the rate of the reverse reaction. The hydrophobic/hydrophilic character of zeolites is associated with their aluminium content [6]. Thus, HUSY is considered a hydrophilic zeolite, because of the total Si/Al ratio around 2.5 (in HUSY, the framework Si/Al ratio is around 5, but there are extra-framework aluminium, formed upon steam dealumination, that contributes to the total Si/Al ratio).

Glycerol acetals were produced by the reaction of glycerol with butanal, pentanal, hexanal, octanal and decanal over Amberlyst-15 acid resin. The conversion decreased with the size of the aldehyde chain, possibly due to aggregation of the hydrocarbon chain to minimize repulsive interactions with the polar environment (reverse micelle). The glycerol acetals were blended with animal fat biodiesel to evaluate its performance as antifreezing agents. The results indicated that addition of 5 vol.% of butanal/glycerol acetal isomers reduces the pour point of the biodiesel in 5°C [7].

Reaction of glycerol with aromatic aldehydes affords a mixture of acetal isomers with antioxidant capability. To perform as antioxidant, a molecule must form a relatively stable-free radical upon hydrogen abstraction. The acetals of glycerol and aromatic aldehydes possess a benzylic C–H bond, which may give rise to a stable-free radical, depending on the substituent of the aromatic ring.

Acetals of glycerol and aromatic aldehydes, such as benzaldehyde, anisaldehyde, p-nitrobenzaldehyde and p-chlorobenzaldehyde, were prepared and subsequently tested in the decay of the diphenyl-picrylhydrazine radical (DPPH•) [8]. This is a traditional test used to assess the antioxidant potential of a substance. The DPPH• radical has a dark violet colour in solution, but upon hydrogen abstraction, it turns into a pale yellow solution (Scheme 5.3). Figure 5.2 shows a qualitative colour test of the glycerol/aromatic aldehydes acetals in the DPPH• test. It can be

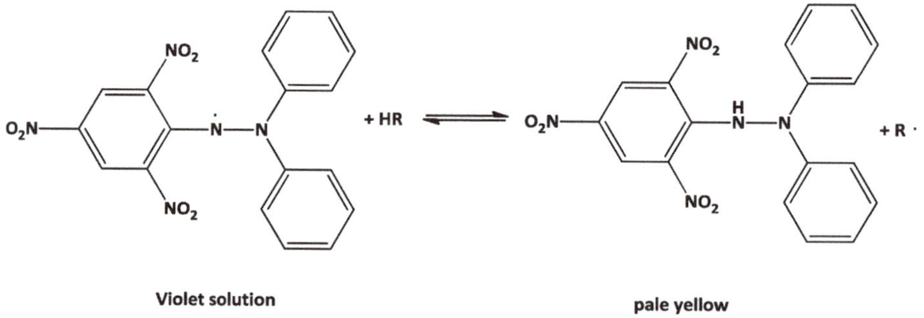

NO_2 O_2N N N NO_2 + HR ⇌ NO_2 O_2N H N N NO_2 + R·

Violet solution pale yellow

Scheme 5.3 DDPH test of antioxidation potential of a substance

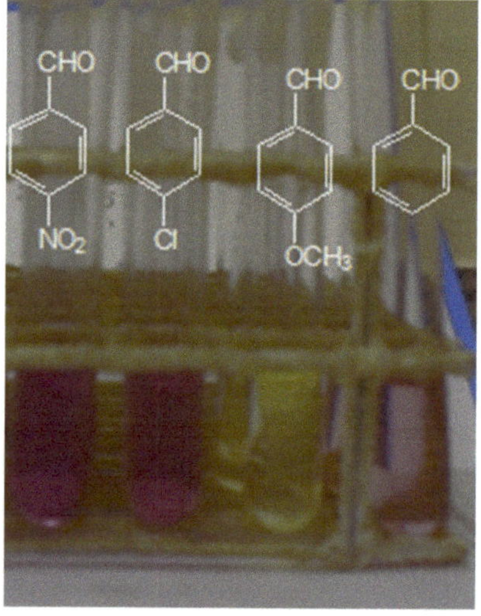

Fig. 5.2 Qualitative DPPH test with the glycerol/aromatic aldehyde acetals (the structures of the aldehydes are shown)

seen that the test solution is more decolorized with the acetals of glycerol and anisaldehyde, whereas it is still violet when the acetals of glycerol and p-nitrobenzaldehyde or p-chlorobenzaldehyde were used. This is because the p-methoxy group can better stabilize the radical by resonance, as shown in Fig. 5.3.

The antiradical activity of the glycerol/anisaldehyde acetal was determined as the amount of antioxidant necessary to decrease the initial DPPH• radical concentration by 50%, EC_{50} (Fig. 5.4). The $EC_{50} = 80$ is similar to coumaric acid or vanillin [8].

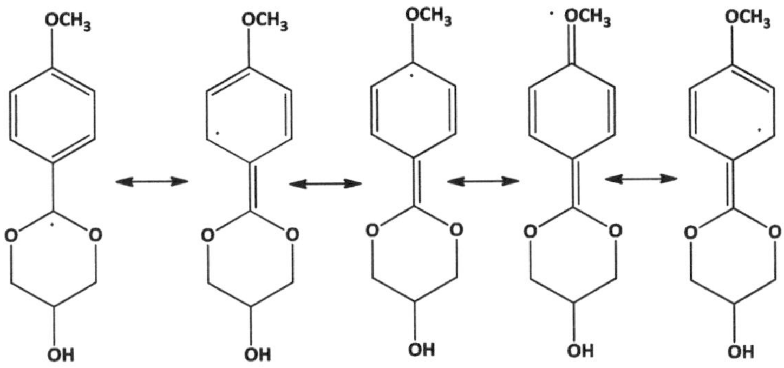

Fig. 5.3 Resonance structures in the glycerol/anisaldehyde acetal, showing delocalization of the electron on the methoxy group

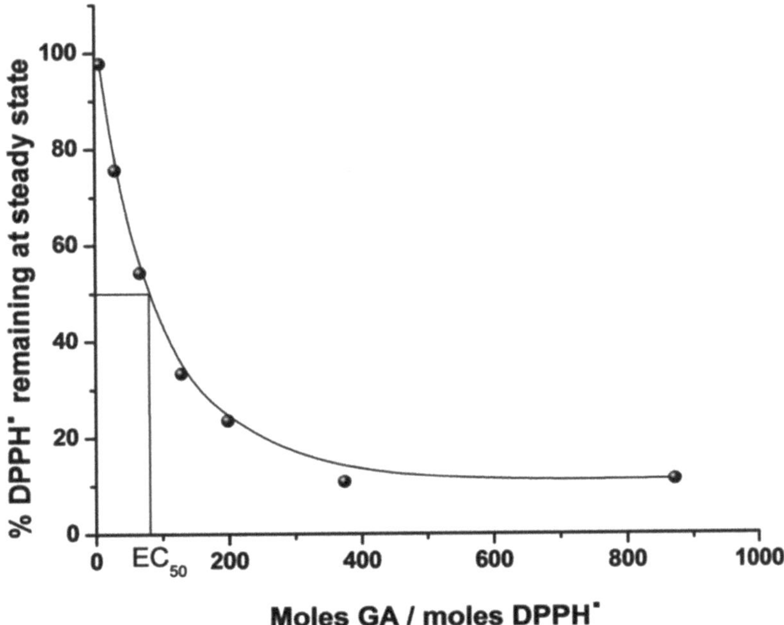

Fig. 5.4 Determination of EC_{50} for glycerol/anisaldehyde acetal, the ratio of GA/DPPH• at which the percentage of DPPH• remaining at the steady state is 50% (Reprinted with the permission of author(s), "Glycerol Acetals with Antioxidant Properties"; in: J. R. Dodson, T. Avellar, J. Athayde, Claudio J. A. Mota, Pure & Applied Chemistry, 86, De Gruyter, 2016, pp. 905–912, Fig. 3)

The glycerol formal, produced upon the reaction of glycerol with formaldehyde, finds applications as a disinfectant and solvent for cosmetics and medical usage. Glycerol acetals are used as hydrophobic intermediates in non-ionic surfactant synthesis [9].

5.2.2 Glycerol Ketals Through the Reaction with Ketones

The ketalization of glycerol with acetone was studied with various acid catalysts such as Amberlyst-15, K-10 Montmorillonite, zeolite H-Beta, HUSY, zeolite HZSM-5 and p-toluenesulfonic acid [4]. The reaction proceeds faster than with formaldehyde and Amberlyst-15 sulfonic acid resin presented around 95% conversion after 40 min (Fig. 5.5). K-10 Montmorillonite and zeolite Beta also show high conversions, with zeolites HUSY and HZSM-5 showing the worst performance. In the case of HZSM-5, the explanation is due to the pore constraint system that favours shape selectivity properties. In the case of HUSY, the poor activity is associated with the hydrophilic character of this zeolite.

Solketal was blended in 1, 3 and 5 vol% with gasoline containing 0 and 25 vol% of ethanol. The addition of solketal did not significantly change the distillation curve of the gasolines but reduced gum formation and increased the octane number up to 2.5 points (Figs. 5.6 and 5.7). Hence, solketal can be used as a good gasoline additive, improving octane number and reducing gum formation, which is especially

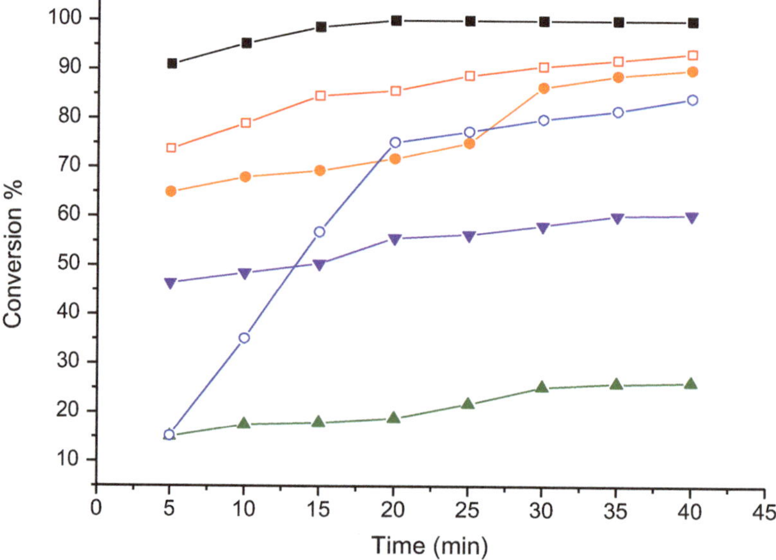

Fig. 5.5 Kinetics of the glycerol reaction with acetone at 70°C over various acid catalysts. (black-filled square) Amberlyst-15, (*red open square*) zeolite Beta, (*orange-filled circle*) K-10, (*blue open circle*) PTSA, (*violet-filled inverted triangle*) HUSY and (*green-filled triangle*) zeolite ZSM-5 (Reprinted from Ref. [4] with permission of the RSC)

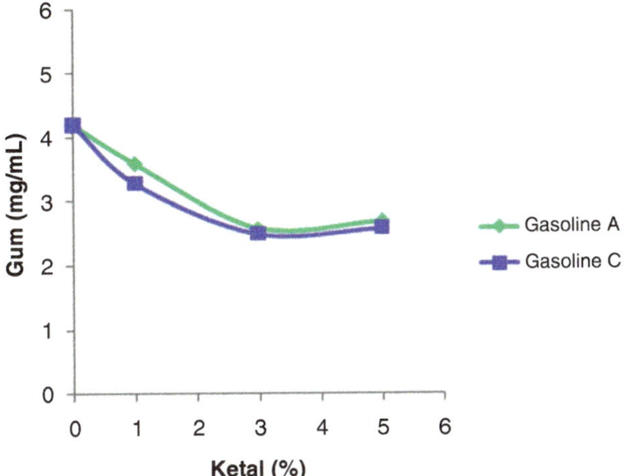

Fig. 5.6 Gum formation against the volume percent of solketal. Gasoline A has no ethanol, whereas gasoline C has 25 vol% of ethanol

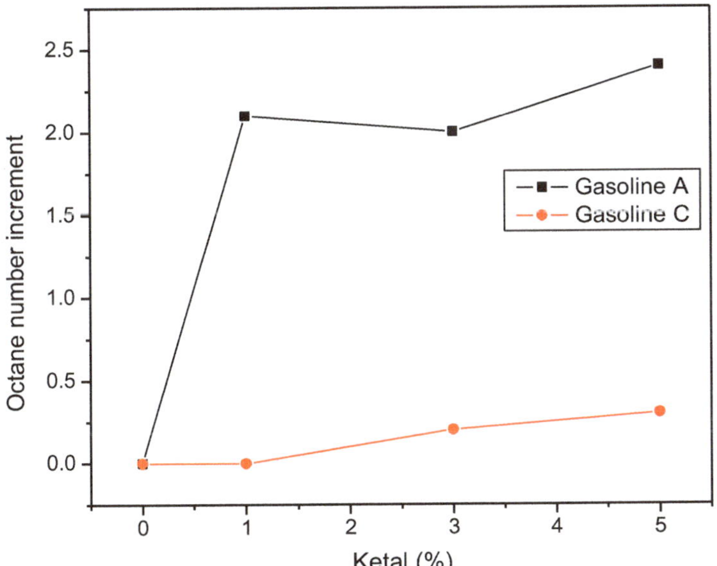

Fig. 5.7 Octane number increment (average of MON and RON) against the volume percent of solketal. Gasoline A has no ethanol, whereas gasoline C has 25 vol% of ethanol

important in gasolines produced from catalytic cracking, where the concentration of olefins is high [10].

Methyl *tert*-butyl ether (MTBE) was the main oxygenated additive used in gasoline up to the beginning of this century. It is produced from the acid-catalysed reaction of methanol with isobutene. MTBE has been phased out because of its carcinogenic properties [11]. Upon leaking of fuel from the storage tanks in gas stations, MTBE may be accumulated in water sources and ultimately ingested by humans. Therefore, its persistency in water, together with the potential mutagenic power, has contributed to the discontinued use in gasolines. On the other hand, solketal is easily decomposed in water. The reactivity of solketal towards acid-catalysed hydrolysis was studied in relation to temperature, molar ratio and catalyst loading. At 80°C, 5:1 water/ketal molar ratio and 3.0 mmol of Amberlyst-15 sulfonic acid resin, the hydrolysis of solketal was almost complete, indicating that it may be decomposed in excess water [12].

Most of the studies on glycerol acetalization and ketalization were carried out with pure glycerol. Nevertheless, the glycerine that comes from biodiesel plants has many impurities, which may impair its proper conversion into high-added value compounds. The acid-catalysed reaction of glycerol with acetone in the presence of some common impurities that might be present in the crude glycerine of biodiesel production has been investigated [13]. Methanol is present in minor amounts and comes from an incomplete removal by distillation after the process. Water may be present due to washing procedures, whereas sodium chloride is formed upon the neutralization of the basic catalyst with HCl solutions. Typically, a crude glycerine from a Brazilian biodiesel plant has around 80% of glycerol, 7–10% of water, up to 1% methanol and around 10–15% of NaCl. Table 5.1 shows the results of glycerol conversion to solketal, as a function of the contaminant, using Amberlyst-15 acid resin as catalyst. The conversion of a crude glycerol from a Brazilian biodiesel plant is also shown. It can be seen that with the crude glycerol, the conversion is significantly lower compared with the pure glycerol. The decrease in catalytic activity is mainly caused by the simultaneous presence of water and NaCl, which neutralizes the acid sites. Nevertheless, high water concentration (above 15%) is also deleterious, because it weakens the acid sites and probably favours the reverse reaction [13].

Solketal can be used as a cold flow additive to enhance the properties of the diesel fuel. It also reduces the fuel viscosity [14].

Table 5.2 exhibits the main different types of catalysts used in the glycerol acetals and ketals synthesis.

Table 5.1 Conversion of glycerol in the reaction with acetone at 70°C and in the presence of impurities. Amberlyst-15 was used as acid catalyst

Glycerol	Methanol (%)	Water (wt%)	NaCl (wt%)	Conversion (%)
Pure	–	–	–	95
Pure + methanol	1	–	–	76
Pure + methanol	5	–	–	78
Pure + water	–	5	–	79
Pure + water	–	10	–	65
Pure + water	–	15	–	39
Pure + NaCl	–	–	5	71
Pure + NaCl	–	–	15	71
Pure + methanol + water	1	10	–	73
Pure + water + NaCl	–	10	5	53
Pure + water + NaCl	1	10	5	47
Crude glycerine	0.2	7	12.8	50

Table 5.2 Selected results of glycerol reaction with acetone and aldehydes

Catalyst	Reactions conditions	Results[a]	Reference
Amberlyst-15	Glycerol/acetone (1:2); 70°C, 45 min	C = 100%	[4]
Amberlyst-15	Glycerol/formaldehyde (1:2); 70°C, 45 min	C = 60%	[4]
Amberlyst-15	Glycerol/acetone 1:6; 70°C, 30 min	C = 85%	[15]
Amberlyst-36	Glycerol/acetone 1:4; 25°C, WHSV 2 h^{-1}	Y = 94%	[16]
Amberlyst-15	Glycerol/butanal 1:1.05; 70°C, 2 h	C = 78% S = 80% (six-membered acetal)	[7]
Ar-SBA-15[b]	Glycerol/acetone 1:6; 70°C, 30 min	C = 82%	[15]
H-Beta	Glycerol/acetone 1:2; 28°C, 1 h	C = 86% S = 98% (solketal)	[17]
H-Beta	Glycerol/acetone 1:2; 70°C, 45 min	C = 92%	[4]
H-Beta	Glycerol/ formaldehyde 1:2; 70°C, 45 min	C = 98%	[4]
HUSY	Glycerol/acetone 1:1; 80°C, 6 h	C = 36%	[18]
K-10	Glycerol/acetone 1:2; 70°C, 45 min	C = 90%	[4]
K-10	Glycerol/ formaldehyde 1:2; 70°C, 45 min	C = 62%	[4]
K-10	Glycerol/benzaldehyde 1:1.2; room temperature; 6 h	Y = 95%	[3]
Ni/AC[c]	Glycerol/acetone 1:8; 45°C, 3 h	C = 98% S = 86% (solketal)	[19]
Zr/AC[c]	Glycerol/acetone 1:8; 45°C, 3 h	C = 67% S = 63% (solketal)	[19]
Ni-Zr/AC[c]	Glycerol/acetone 1:8; 45°C, 3 h	C = 100% S = 74% (solketal)	[19]
SiO_2	Glycerol/benzaldehyde 1:1; 100°C, 8 h	C = 23% S = 50% (five-membered acetal)	[20]
MoO_3/SiO_2	Glycerol/benzaldehyde 1:1; 100°C, 8 h	C = 72% S = 40% (five-membered acetal)	[20]
MoPO/SBA-15	Glycerol/acetone 1:2; room temperature, 2 h	C = 100% S = 98% (solketal)	[21]
MoO_x/TiO_2-ZrO_2	Glycerol/benzaldehyde 1:1; 100°C, 30 min	C = 78% S = 50%	[22]
MoO_x/ZrO_2	Glycerol/acetone 1:6; room temperature; 90 min	C = 88% S = 97% (solketal)	[23]

[a] *C* conversion, *Y* yield, *S* selectivity
[b] Ar-SBA-15, arenesulfonic acid-functionalized mesostructured silica
[c] *AC* activated carbon

5.3 Glycerol Ethers

5.3.1 Etherification of Glycerol

Transformation of glycerol into glycerol ethers is a good way to decrease the polarity, viscosity and boiling point, thus allowing glycerol to be incorporated in hydrocarbon-based fuels as oxygenated additives [24].

There are basically three main routes of glycerol etherification: acid-catalysed reaction with olefins, like isobutene; acid-catalysed reaction with alcohols, which also includes the self-etherification of glycerol to yield dimers and trimers; and the reaction of glycerol alkoxides with alkyl halides or alkyl sulphates. This later route is less studied in the literature, because it involves expensive and hazardous reactants. Nevertheless, it can afford the triglyceryl alkyl ethers in high yield, as shown in the reaction of glycerol with dimethyl sulphate in the presence of NaOH or KOH [25].

5.3.2 Etherification of Glycerol with Isobutene

Glycerol etherification using isobutene may be carried out in a batch reactor, under heating and pressure, in the presence of a suitable acidic catalyst. Ethers of glycerol are formed as a result of three consecutive steps, such as shown in Scheme 5.4.

Two side reactions give rise to the formation of *tert*-butyl alcohol (TBA) and isobutene oligomers, chiefly dimers (DIB) (Scheme 5.5). In fact, the reaction of glycerol with isobutene is similar to the reaction of methanol with this same olefin to obtain MTBE. The reaction proceeds with the protonation of the olefin to form the *tert*-butyl cation, which is then nucleophilically attacked by the alcohol, followed by deprotonation to afford the ether.

The glycerol etherification with isobutene has been extensively investigated. Among the catalyst used are sulfonic mesostructured silicas, sulfonic acid resins and zeolites. Many catalyst give complete glycerol conversion with selectivity of 90% to the ethers [26].

The reaction can also be performed with *tert*-butyl alcohol, affording, basically, the same products as in the case of isobutene. The main products obtained in glycerol etherification using TBA are 3-*tert*-butoxy-1,2-propanediol or 2-*tert*-butoxy-1,3-propanediol (MTBG), 2,3-di-*tert*-butoxy-1-propanol or 1,3-di-*tert*-butoxy-2-propanol (DTBG) and 1,2,3-tri-*tert*-butoxy-propane (TTBG), as shown in Scheme 5.6. However, water is formed as by-product and may interact with the catalyst, weakening the acid strength.

The *tert*-butyl glycerol ethers can be used as oxygenated fuel additives, replacing methyl *tert*-butyl ether (MTBE) in gasoline. The MTBG ethers have low solubility in hydrocarbons, but DTBG and TTBG are soluble in diesel, making them good additives to this fuel. They have high potential for blending with diesel and biodiesel formulations, enhancing the combustion efficiency and decreasing the emissions of pollutant gases [27].

$$\mathrm{G + IB \rightleftharpoons MTBG} \quad (1)$$
$$\mathrm{MTBG + IB \rightleftharpoons DTBG} \quad (2)$$
$$\mathrm{DTBG + IB \rightleftharpoons TTBG} \quad (3)$$

Scheme 5.4 Glycerol (G) etherification with isobutene (I) to form mono-*tert*-butyl glycerol ethers (MTBG), di-*tert*-butyl glycerol ethers (DTGB) and tri-*tert*-butyl glycerol ethers (TTBG)

Scheme 5.5 Formation of TBA and DIB

$$\mathrm{IB + H_2O \rightleftharpoons TBA}$$
$$\mathrm{2\,IB \longrightarrow DIB}$$

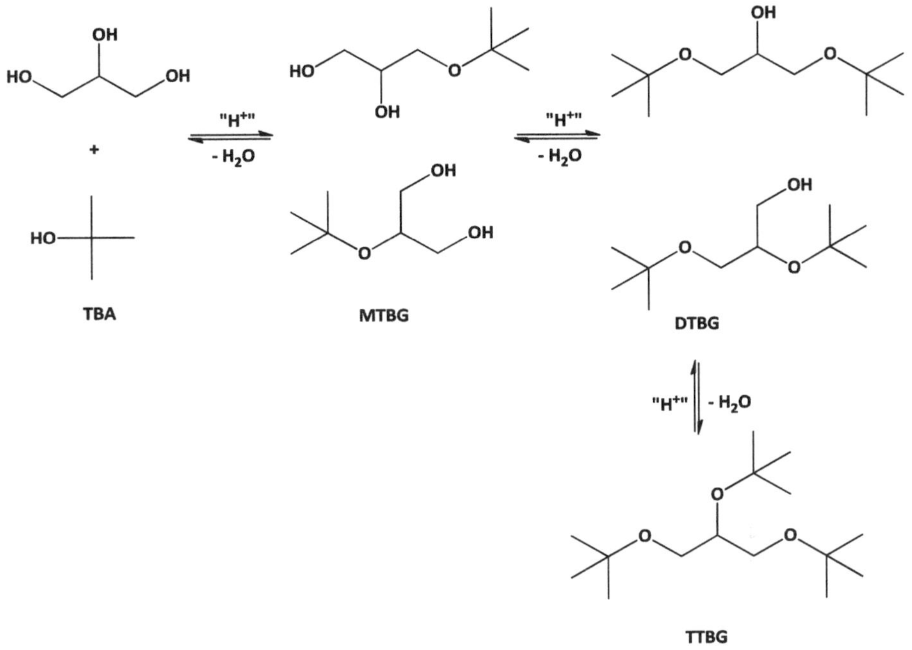

Scheme 5.6 Etherification of glycerol with *tert*-butyl alcohol

5.3.3 *Etherification of Glycerol with Ethanol*

Isobutene is primarily derived from fossil sources, and the *tert*-butyl glyceryl ethers discussed previously are not completely renewable fuel additives. On the other hand, ethanol is obtained from biomass, mainly from the fermentation of sugars, being commonly used as gasoline additive.

The etherification of glycerol with ethanol occurs in the presence of an acidic catalyst (Scheme 5.7). The primary OH groups react preferentially, probably due to steric problems. The reaction can be reversible, and the water formed in each step competes with the reactants for the active sites, decreasing the glycerol conversion and formation of more substituted ethers.

Different types of acidic heterogeneous catalysts, including sulfonic acid resins, zeolites and grafted silicas were used for the synthesis of mono-ethers of glycerol and ethanol. For instance, with Amberlyst resin, monomethyl glyceryl ethers were selectively produced up to a glycerol conversion of 40% at 160°C. The best results were found at 200°C, using grafted silica, when 68% glycerol conversion and 75% selectivity to mono and 25% selectivity to the diethyl ethers were achieved. Silicon-rich zeolites (Si/Al = 25) also present good activities, showing glycerol conversion of 57% and selectivity of 75% and 25% to the mono and diethyl glyceryl ethers, respectively [26].

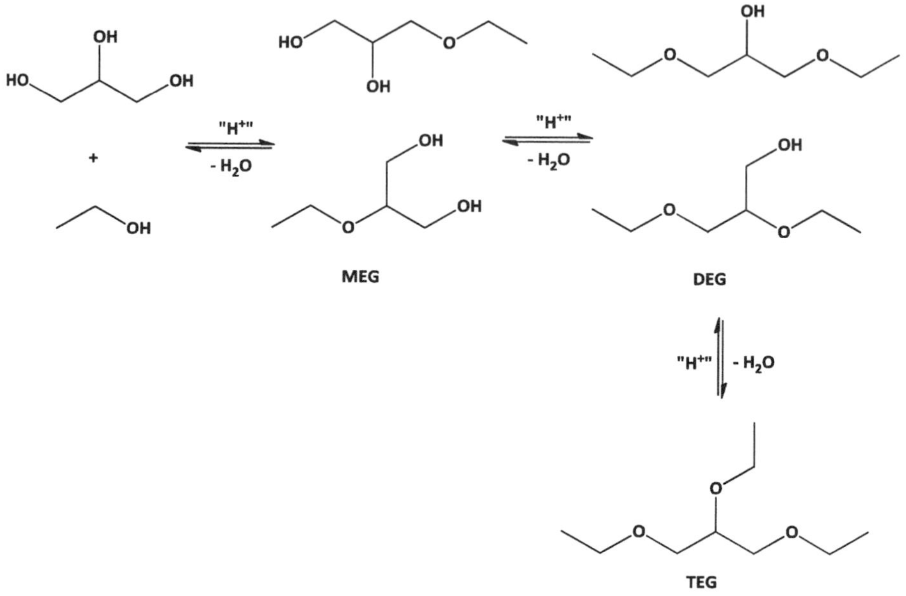

Scheme 5.7 Acid-catalysed etherification of glycerol with ethanol

The etherification of glycerol with bioethanol was studied using different acid catalysts, such as HZSM5, H-Beta, tungstophosphoric acid, $FeCl_3$, $AlCl_3$ and H_2SO_4. Tungstophosphoric acid exhibited 97% conversion of glycerol at 160°C, ethanol/glycerol molar ratio of 6:1 and 20 h. The selectivity to the mono-ethyl ether was 62% but decreased with increasing conversion along with the increase of the selectivity towards di (28%) and tri ethers (10%) [28].

The etherification of biodiesel-derived glycerol with anhydrous ethanol over arenesulfonic acid-functionalized mesostructured silicas was studied to produce ethyl ethers of glycerol. The best conditions were 200°C, molar ratio of ethanol to glycerol of 15:1 and catalyst loading of 19 wt%. A glycerol conversion of 74% was observed, with 42% yield of the ethyl ethers after 4 h. However, there was significant formation of by-products [29].

The etherification of glycerol over H-Beta zeolite catalyst presented 85% conversion after 4 h (molar ratio 1:3) and 180°C. The main product formed was the mono-ethyl glycerol ethers with a selectivity of 74%. The di- and tri-ethyl glyceryl ethers were also produced but with selectivity of 15% and 11%, respectively. The increase in molar ratio leads to a decrease in conversion, because of the competitive self-etherification of ethanol yielding diethyl ether.

The di- and triglyceryl ethers have potential as additives for biodiesel and diesel formulations, because of the low polarity and miscibility with water. The mono-ethyl ether is soluble in water and may be an interesting intermediate for the production of other chemicals [28].

Alkyl glycerol ethers are potential fuel additives that may reduce the emissions of particulate matters, carbon monoxide and carbonyl compounds in exhaust gases. They can decrease the cloud point of diesel fuel when combined with biodiesel. The appropriate ratio of the glycerol ethers, alcohol and gasoline could reduce the vapour pressure to the desired level and reduce fuel consumption. The presence of glycerol ethers could help in decreasing the gel temperature of fuels, which leads to the reduction of the viscosity. The presence of hydroxyl groups could also lower the NOx emission [30, 31].

A limitation on the use of biodiesel in some regions is the cloud point. Petroleum-derived diesel fuels present cloud point around −16°C, but soybean biodiesel, for instance, shows cloud point around 0°C. The addition of the ethers can decrease the cloud point of diesel fuels [32].

A mixture of mono- and diethyl glyceryl ethers was blended with soybean and tallow biodiesel formulations. The blends showed a decrease of the cloud point in 2 and 4°C for soybean and tallow biodiesel, respectively, with respect to the neat biodiesel formulations. The reduction in the pour point was even higher, being 5 and 3°C for soybean and tallow biodiesel, respectively [33].

5.3.4 Oligomerization of Glycerol

Glycerol dimers are commercially produced by the base hydrolysis of epichlorohydrin using NaOH or Na_2CO_3 solution [34]. Epichlorohydrin may be obtained from the reaction between glycerol and HCl in the Epicerol® process, as highlighted in Chap. 4.

Another route of production of polyglycerols is the self-etherification at elevated temperature under acid or base catalysis. This one-step process avoids hazardous or highly reactive intermediates, such as epichlorohydrin.

Homogeneous and heterogeneous catalysts have been investigated in the glycerol self-etherification. Homogeneous catalysts, such as H_2SO_4, NaOH, Na_2CO_3, CsOH and Cs_2CO_3, lead to higher conversions, but lower selectivity to the di- and triglycerol oligomers, when compared with the heterogeneous catalysts [35].

The basicity and solubility of the catalysts in glycerol have a great influence in the yield and selectivity of the reaction. Alkali hydroxides are strong bases but less soluble in glycerol than carbonates. Heterogeneous catalysts can be separated from the reaction mixture and sometimes may be reused. Nevertheless, they present high cost, low thermal stability and low surface area. Moreover, higher temperatures and longer reaction times are often applied when using heterogeneous catalysts [36, 37].

The glycerol conversion to diglycerol yields three configurational isomers (Scheme 5.8).

The main challenge in the production of diglycerols consists in finding a selective route, preventing or minimizing the formation of acrolein, cyclic compounds and higher oligomers. Batch conversion processes demand fractional distillation to separate the glycerol, diglycerols and the higher oligomers [38].

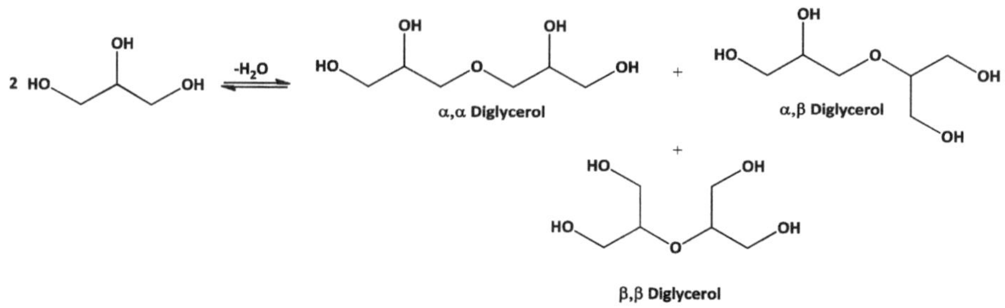

Scheme 5.8 Glycerol dimerization and configurational isomers of diglycerol

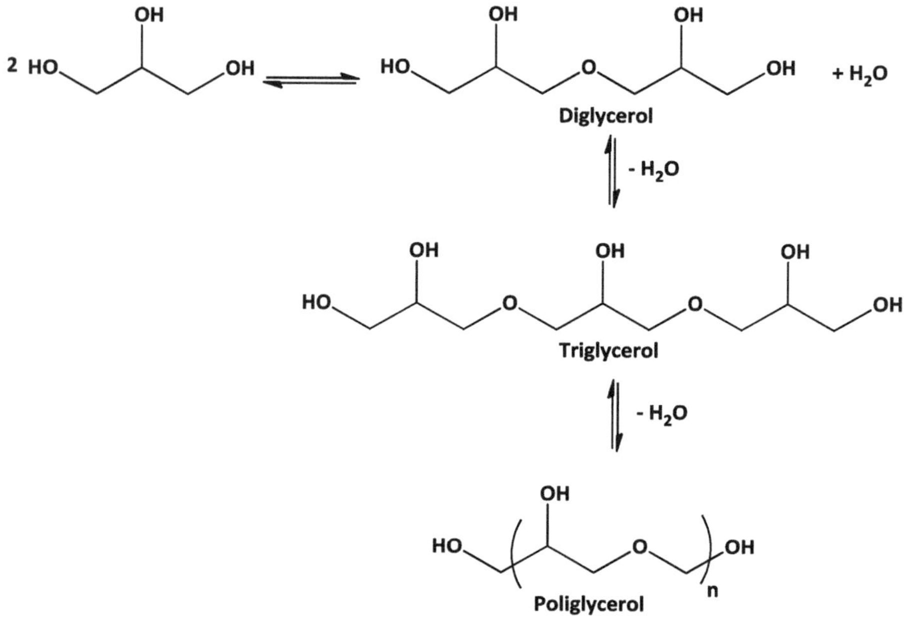

Scheme 5.9 Self-etherification of glycerol to form polyglycerols

The self-etherification of glycerol to form polyglycerols is shown in Scheme 5.9.

The selective synthesis of di- and triglycerol can be carried out over mesoporous solids and basic materials with yields over 80% [39].

Self-etherification of glycerol over solid bases has been studied in a discontinuous batch reaction, using Cs-exchanged ZSM-5 at 260°C during 24 h. Glycerol conversion was 80% with selectivity to di- and triglycerol of 90% [40]. MCM-41 catalysts impregnated with different amounts of Cs showed 85% glycerol conversion after 8 h at 260°C [41].

Polyglycerols are used as surfactants and lubricants. Di- and triglycerols are starting materials for the production of emulsifiers in the food and cosmetic industries or as additives in polyvinyl chloride manufacture [32].

When $CsHCO_3$ was used as catalyst, 100% of linear diglycerol is produced at 10% glycerol conversion. The α,α diglycerol isomer was preferentially formed because the reaction is governed by the probability that two terminal OH groups of glycerol can react with two terminal OH groups of a second glycerol molecule [42].

Glycerol self-etherification was produced with sodium carbonate as catalyst. Generally, metal carbonate catalysts are more active than hydroxides because of their higher solubility in glycerol [34].

The selective self-etherification of glycerol to di- and triglycerol was studied in the presence of alkaline earth metal oxides. The glycerol conversion increased with increasing catalyst basicity. Selectivity of over 90% to di- and triglycerols was observed with CaO, SrO and BaO, at 60% conversion [43].

5.4 Glycerol Esters

5.4.1 Esterification with Acetic Acid and Acetic Anhydride: Formation of Acetins

The products of mono-, di- and tri-acetylation of glycerol, respectively, monoacetin, diacetin and triacetin have great industrial applications. The mono- and diacetylated esters may be used in cryogenics and as raw materials for the production of biodegradable polyesters. Triacetin can be used as biocide, plasticizer, solvent in cosmetic formulations and also as additive to biodiesel. Another use is as carrier for flavours and fragrances, being generally recognized as a safe human food ingredient by the Food and Drug Administration of the USA [44, 45].

The acetylation of glycerol with acetic acid has been studied using different solid acid catalysts at 120°C (Scheme 5.10). Amberlyst-15 acid resin presented the best performance; after 30 min of reaction time, the glycerol conversion was 97%, with a selectivity of 54% to diacetin, 31% to monoacetin and 13% to triacetin. Zeolites HZSM-5 and HUSY presented the worst performance, probably due to diffusion and acid site deactivation problems due to the formation of water. In all cases hydroxy acetone, originated from glycerol dehydration, was formed in minor amounts [46].

The acetylation of glycerol with acetic anhydride occurs at mild reaction conditions. With zeolite H-Beta and K-10 Montmorillonite as catalysts, the yield of triacetin was 100% at 60°C within 20 min. The uncatalysed reaction showed 10% of monoacetin, 54% of diacetin and 34% of triacetin at the same temperature but after 120 min (Scheme 5.11). It is interesting to note that H-Beta was highly active when acetic anhydride was used as the acetylating agent, yielding 100% of triacetin at 60°C and 20 min. Nevertheless, when acetic acid was employed as acetylating agent, the performance of zeolite H-Beta was basically the same of the blank, uncatalysed reaction (Table 5.3), with a similar distribution of the acetin products [47].

Acetic anhydride is known to be more reactive than acetic acid, and this explains the higher selectivity to triacetin at milder reaction conditions. When acetic acid is used, the reaction is reversible and water is formed as by-product. Notwithstanding, zeolite H-Beta is a hydrophobic zeolite and performs well in the presence of water.

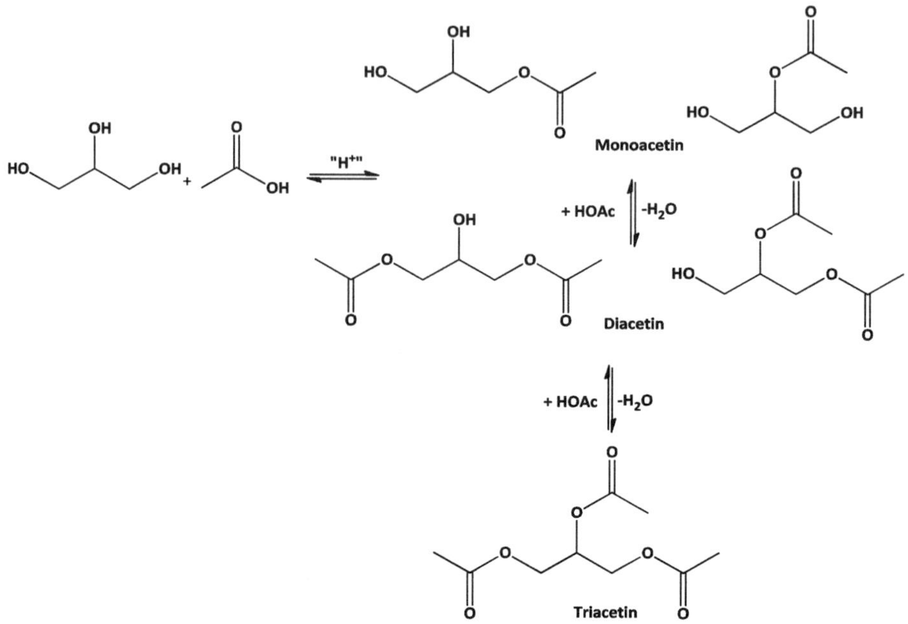

"H+"

Monoacetin

+ HOAc $-H_2O$

Diacetin

+ HOAc $-H_2O$

Triacetin

Scheme 5.10 Acid-catalysed acetylation of glycerol with acetic acid

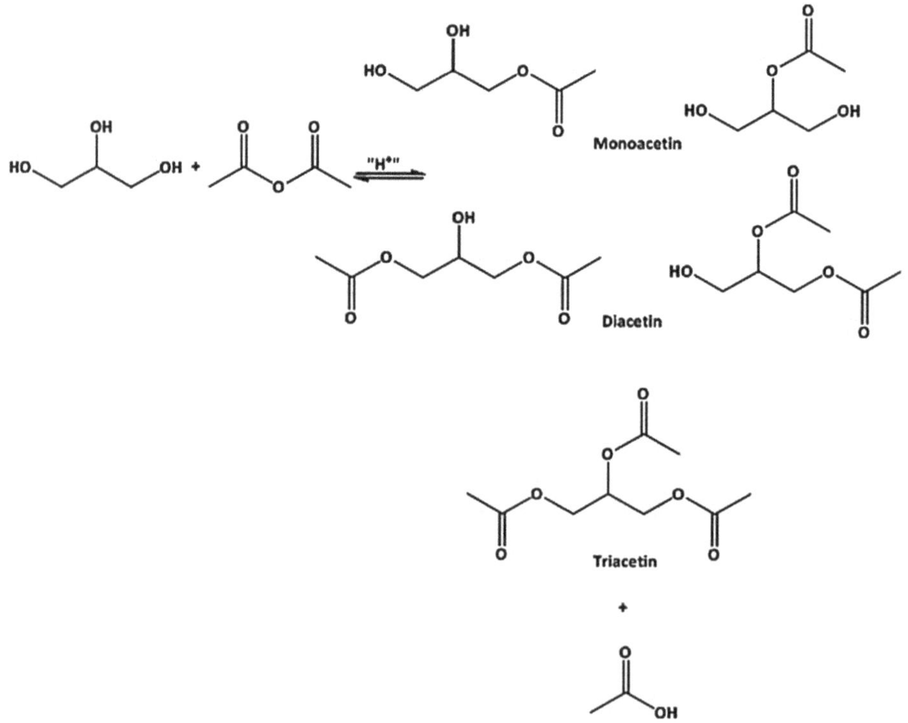

"H+"

Monoacetin

Diacetin

Triacetin

+

Scheme 5.11 Acetylation of glycerol with anhydride acetic

Table 5.3 Acetylation of glycerol with different catalysts and acylation agents

Catalyst	Acetylating agent (molar ratio)[a]	Temperature (°C)	Time (min)	Selectivity to acetins[b] (%)		
				Mono	Di	Tri
H-Beta	Ac_2O (4:1)	60	20	–	–	100
K-10	Ac_2O (4:1)	60	20	–	–	100
Amberlyst-15	Ac_2O (4:1)	60	80	–	–	100
Blank	Ac_2O (4:1)	60	120	10	56	34
H-Beta	HOAc (4:1)	120	120	48	39	4
K-10	HOAC (4:1)	120	120	36	52	6
Amberlyst-15	HOAc (4:1)	120	120	18	55	24
Blank	HOAc (4:1)	120	120	50	40	4

[a] Ac_2O stands for acetic anhydride and HOAc for acetic acid
[b] Reactions with HOAc also produced hydroxy acetone, which accounts for the remaining distribution difference

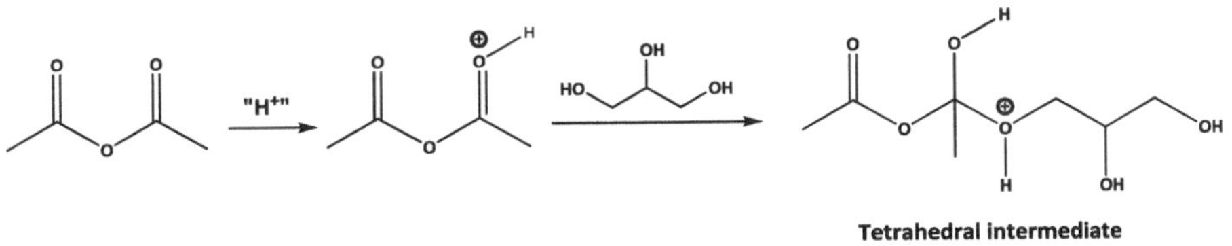

Scheme 5.12 $A_{AC}2$ mechanism for the glycerol acetylation with acetic anhydride. Formation of the tetrahedral intermediate

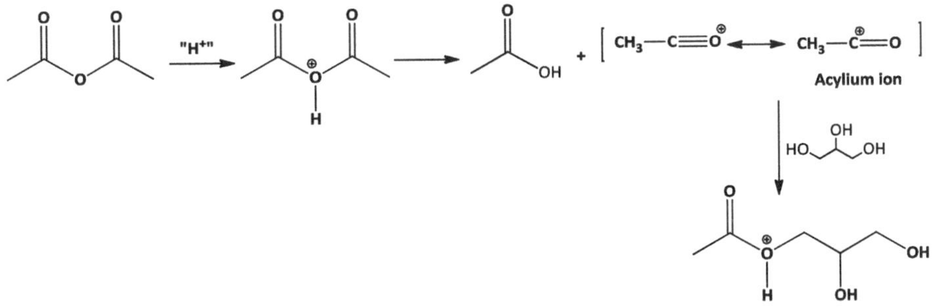

Scheme 5.13 $A_{AC}1$ mechanism for the acetylation of glycerol with acetic anhydride. Formation of the acylium ion

Thus, the explanation for its poor catalytic performance in the acetylation of glycerol with acetic acid cannot be explained in terms of hydrophobicity/hydrophilicity.

There might be two possible mechanisms of esterification in strong acidic medium: the normal $A_{AC}2$ mechanism, involving protonation of the carbonyl oxygen atom and nucleophilic attack to form a tetrahedral intermediate (Scheme 5.12), and the $A_{AC}1$ mechanism, where protonation occurs on the oxygen atom attached between the two carbonyl groups, followed by formation of an acylium ion (Scheme 5.13). The first pathway is normally preferred, but can be inhibited in situations of steric constraints, as it might occur inside the zeolite cages, especially for the formation of di- and triacetylated products. In these cases, the mechanism involving the acylium ion may prevail [48].

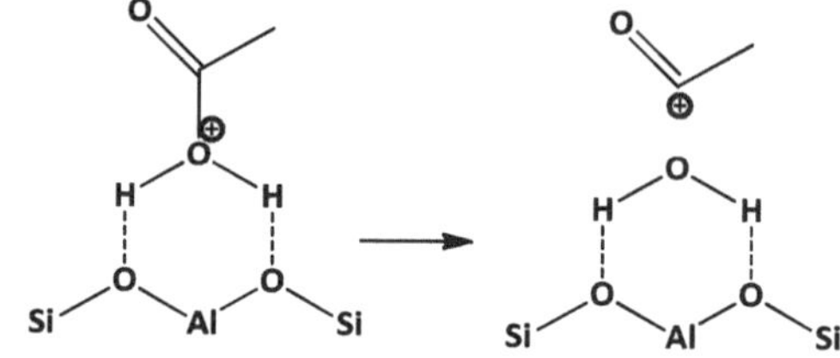

Scheme 5.14 Proposed structure of the protonated acetic acid on the zeolite surface and its dehydration to the acylium ion (no zeolite assistance)

Scheme 5.15 Proposed structure of the protonated acetic anhydride and its transformation to the acylium ion (zeolite assisted)

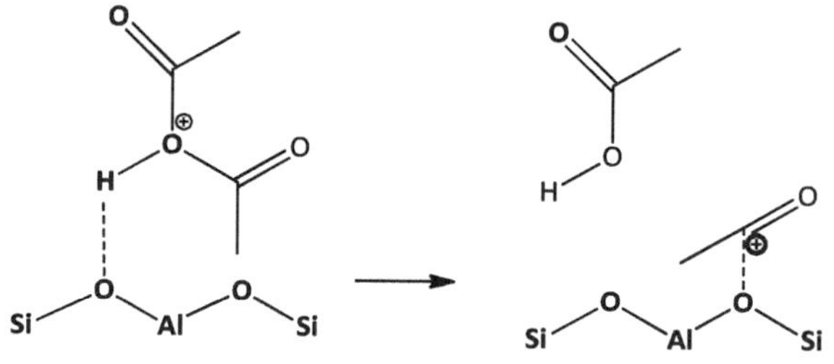

The explanation for the different catalytic behaviour of zeolite H-Beta may be based on the interaction of the protonated acetic acid and protonated acetic anhydride with the zeolite surface [47]. Scheme 5.14 shows a possible structure of the protonated acetic acid ($A_{AC}1$ mechanism), where two hydrogen atoms are interacting with the zeolite structure through hydrogen bonds. In this case, formation of the acylium ion may not be assisted by the zeolite surface increasing the energy of activation. On the other hand, in the case of acetic anhydride, the zeolite may assist the formation of the acylium ion (Scheme 5.15), contributing to the decrease of the energy of activation. This may explain the different behaviours of the two acetylating agents on H-Beta zeolite.

Niobic acid supported on tungstophosphoric acid catalysts were prepared and evaluated in the acetylation of glycerol with acetic acid. About 88% conversion was observed after 30 min of reaction time. A blank experiment, without using the catalyst, showed only 8% conversion. The tungstophosphoric acid gave 26% conversion under similar experimental conditions. Monoacetin was the main product at the beginning of the reaction, but as the reaction proceeds, the selectivity to di- and triacetin increases, which is mainly due to the progressive acetylation of monoacetin. The glycerol conversion is nearly complete within 1 h of reaction time [49].

Porous carbon-based acid catalysts, prepared by sulphonation of carbonized sucrose, were evaluated in the esterification of glycerol with acetic acid. These catalysts have an amorphous porous structure and high thermal stability. Conversions above 99% with selectivity towards triacetin around 50% were observed using 9:1 molar ratio of the reactants, 5% of catalyst loading, 200°C and 4 h. A conventional Amberlyst-15 sulfonic acid resin showed 99% conversion but with a selectivity to triacetin of only 10%. The catalyst activity is practically the same for the materials carbonized at 600°C and 850°C but is increased for the sample carbonized at 400°C. The triacetin formation takes place over sulfonated sites, which are directly related to the porosity of the material [50].

The usually low selectivity to triacetin with the use of acetic acid as the acetylating agent is associated with the thermodynamics of the reaction. The formation of mono- and diacetin is exothermic, whereas formation of glycerol triacetate from acetic acid is endothermic, being less favourable [51]. On the other hand, the formation of triacetin from acetic anhydride is exothermic and highly favoured by thermodynamics. From an industrial point of view, it would be more convenient to begin the glycerol acetylation with acetic acid and a heterogeneous catalyst, achieving the maximum selectivity to di- and triacetin. Then, the reaction should be finalized with acetic anhydride to obtain the highest selectivity to triacetin.

Table 5.4 exhibits different heterogeneous catalysts used in glycerol acetylation with acetic acid or acetic anhydride.

Table 5.4 Selected catalysts and reaction conditions for the acetylation of glycerol with acetic acid and acetic anhydride

Catalyst	Reaction conditions	Results[a]	Reference
Amberlyst-15	Glycerol/acetic acid (1:24); 110°C; 200 bar; 2 h	Y = 41% S = 100% triacetin	[52]
Amberlyst-15	Glycerol/acetic acid (1:8); 110°C; 270 min	C = 97% S = 45% triacetin	[53]
Amberlyst-15	Glycerol/acetic acid (1:2); 70°C; 30 min	C = 97% S = 54% diacetin S = 13% triacetin	[46]
Amberlyst-15	Glycerol/acetic anhydride (1:4); 60°C; 20 min	C = 95% S = 10% diacetin S = 90% triacetin	[47]
Amberlyst-15	Glycerol/acetic acid (1:9); 105°C; 4 h	C = 99% S = 26% triacetin	[54]
TPA/Nb_2O_5	Glycerol/acetic acid (1:5); 120°C; 4 h	C = 98% S = 58% diacetin S = 18% triacetin	[49]
Nb_2O_5	Glycerol/acetic acid (1:2); 70°C; 30 min	C = 30 S = 83% monoacetin	[46]
Nb_2O_5 nH_2O	Glycerol/acetic acid (1:3); 105°C; 3 h	C = 82 S = 29% diacetin S = 1% triacetin	[55]
H-Beta	Glycerol/acetic anhydride 1:3; 60°C; 120 min	C = 98% S = 38% diacetin S = 62% triacetin	[47]
H-Beta	Glycerol/acetic anhydride 1:5; 100°C; 50 min	C = 100% S = 34% diacetin S = 66% triacetin	[56]

(continued)

Table 5.4 (continued)

Catalyst	Reaction conditions	Results[a]	Reference
TPA[b] anchored to MCM-41	Glycerol/acetic acid 1:6; 100°C; 6 h	C = 90% S = 70% diacetin S = 10% monoacetin	[57]
K-10	Glycerol/acetic acid 1:2; 70°C; 30 min	C = 96% S = 49% diacetin S = 5% triacetin	[46]
K-10	Glycerol/acetic acid 1:8; 85°C; 5 h	C = 65% S = 20% diacetin S = 5% triacetin	[58]
HZSM-5	Glycerol/acetic acid 1:2; 70°C; 30 min	C = 30% S = 83% monoacetin S = 10% diacetin	[46]
HZSM-5	Glycerol/acetic anhydride 1:5; 80°C; 30 min	C = 99% S = 62% diacetin S = 24% triacetin	[59]
HZSM-5	Glycerol/acetic anhydride 1:5; 100°C; 50 min	C = 100% S = 40% diacetin S = 60% triacetin	[56]
Ar-SBA-15	Glycerol/acetic acid 1:9; 125°C; 4 h	C = 92% S = 40% diacetin S = 20% triacetin	[60]
Pr-SBA-15	Glycerol/acetic acid 1:9; 125°C; 4 h	C = 78% S = 52% diacetin S = 28% triacetin	[61]

[a]*C* conversion, *Y* yield, *S* selectivity
[b]*TPA* 12-tungstophosphoric acid

5.4.2 Glycerol Ketal/Acetal Esters

The acylation of the hydroxyl group of solketal affords the solketal acetate and can be carried out with acetic anhydride. Triacetin was formed as by-product, even though highly pure acetic anhydride is used. The solketal acetate and triacetin mixture were blended with biodiesel increasing the flash point of the fuel. Nevertheless, ketal ester does not improve the cold flow properties as much as triacetin does. The viscosity of the biodiesel was also increased upon the addition of solketal acetate and triacetin mixture [44].

The acetylation of solketal and glycerol formal (Scheme 5.16) with acetic anhydride or acetic acid was studied using 1:1 or 1:3 molar ratio of the ketal or acetal to the acetylating agent [62]. Niobium phosphate, Amberlyst-15, K-10 Montmorillonite and zeolite H-Beta were used as catalyst. Figure 5.8 shows the

Scheme 5.16 Acetylation of solketal and glycerol formal

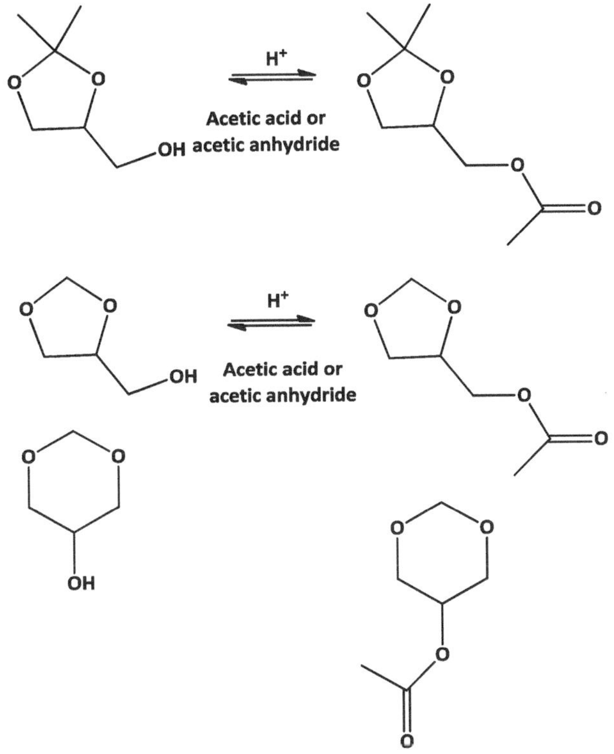

conversion and selectivity to the ketal or acetal esters. The use of acetic anhydride usually led to higher conversions, but the selectivity to the solketal acetate is lower, and triacetin becomes the major product, especially at higher molar ratios of the anhydride to the ketal. It has been shown that triacetin is formed at the expenses of the solketal acetate when an excess of acetic anhydride is present in the medium. The reaction involves the ring opening of solketal acetate by the acylium cations, formed upon the interaction of acetic anhydride with the acidic catalysts, as shown in Scheme 5.17.

The reaction with acetic acid can be efficient. For instance, around 90% conversion and selectivity to solketal acetate were observed with Amberlyst-15 acid resin as catalyst, using 1:3 molar ratio of the ketal to the acid. For the glycerol formal acetals, the conversion was around 60% with a selectivity to the desired acetal esters of only 40%. There occurs hydrolysis of the acetal by the water produced in the esterification, and the glycerol formed may react with the free hydroxyl group of the acetals to afford acetal ethers. Monoacetin is another observed product, formed upon the reaction of glycerol with acetic acid.

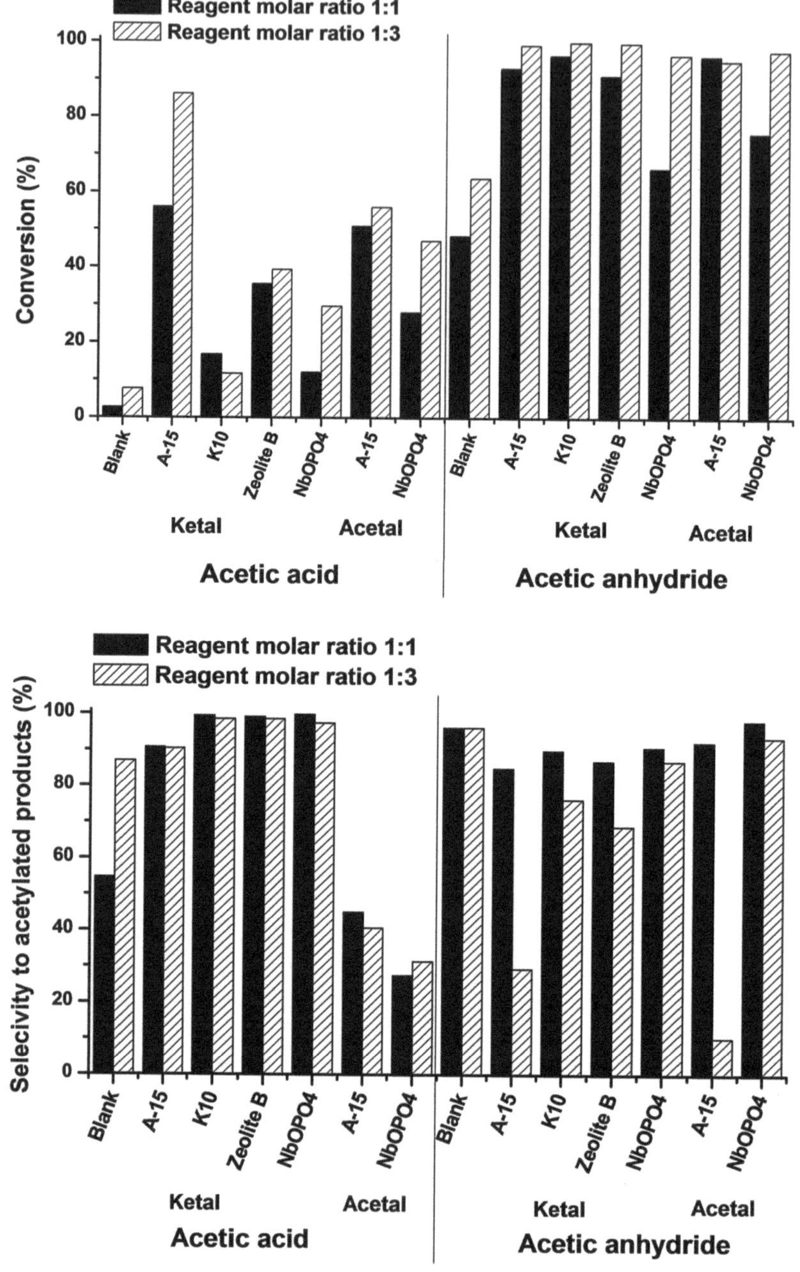

Fig. 5.8 Conversion and selectivity to the respective acetate in the reaction of solketal and glycerol formal acetals with acetic acid and acetic anhydride at different catalysts (Reprinted from Ref. [62] with the permission of Wiley & Sons)

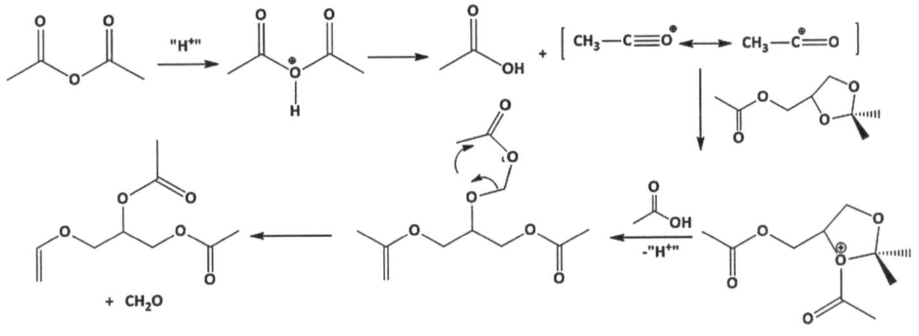

Scheme 5.17 Proposed mechanistic pathway for the formation of triacetin from solketal acetate

5.5 Glycerol Carbonate

Organic carbonates are important chemicals that find applications as monomers for the production of polycarbonates, green solvents, fuel additives and alkylating agents, among others. Glycerol carbonate (GC) is a relatively new compound that is finding new applications every year. Presently, it is still an expensive product, with price higher than US$ 8.00 per kg, which directs applications to specialty and fine chemicals. Among the uses of GC are:

- Electrolyte in lithium ion batteries
- Production of special polymers, like polyurethane plastic coatings
- Emollient, wetting agent and solvent in cosmetics and personal care
- Solvent and carrier in pharmaceutical formulations
- Curing agent in cement and concrete
- Bio-lubricant, intermediate and solvent in the chemical industry

The direct carbonation of glycerol with CO_2 is not thermodynamically favoured at room temperature and atmospheric pressure (Scheme 5.18), but equilibrium is more favoured at high pressures [63].

Indirect methods of production of GC have also been developed; among them are transesterification of glycerol with other organic carbonates and reaction of glycerol with urea. Phosgene can also react with glycerol to afford GC, but this route is less studied because of the high toxicity of phosgene.

5.5.1 *Glycerol Carbonate Through Transesterification*

Dimethyl carbonate (DMC), ethylene carbonate (EC) and propylene carbonate (PC) are normally used in transesterifications with glycerol to produce glycerol carbonate (GC) [64]. The reaction is usually carried out with molar excess of the organic carbonate and in the presence of a basic catalyst, although acid and enzymatic catalysis has also been reported.

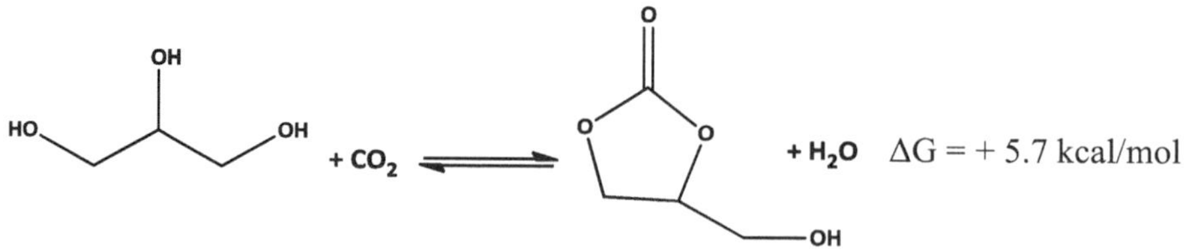

Scheme 5.18 Thermodynamics of the direct carbonation of glycerol with CO_2 at 25°C and 1 atm

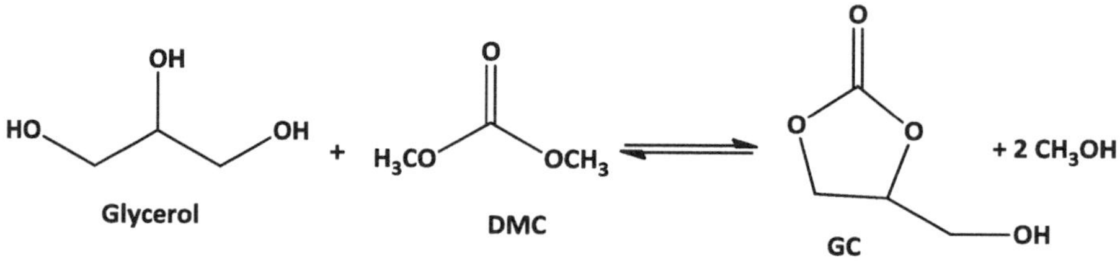

Scheme 5.19 Transesterification of dimethyl carbonate with glycerol

Almost quantitative production of GC can be achieved through transesterification of DMC (Scheme 5.19) with glycerol at temperatures around 70°C and 3–5 h of reaction time, with the use of K_2CO_3 as catalyst. A molar excess of DMC to glycerol of 3:1 to 5:1, together with removal of the formed methanol by distillation, helps in shifting the equilibrium towards GC.

Heterogeneous basic catalysts can also be used, but the conditions are usually more severe, requiring higher temperatures, longer reaction times and higher catalyst loadings in comparison with homogeneous system. For instance, Mg/Al hydrotalcites were employed in the transesterification of DMC with glycerol at 100°C. With a molar ratio of DMC to glycerol of 5:1, 82% yield of GC was obtained after 3 h of reaction time.

A possible by-product formed in the transesterification of organic carbonates with glycerol under base catalysis is glycidol. Usually, higher temperatures lead to the decrease in the selectivity of GC and favours glycidol, which is formed through the base-catalysed decomposition of glycerol carbonate (Scheme 5.20).

Acid catalysis gives poor yields and is not commonly used. Some lipase enzymes can lead to high glycerol conversion and GC yield [65]. For instance, resin-immobilized *Candida antarctica* lipase B may produce glycerol carbonate in 96% yield after 48 h, by the reaction of DMC and glycerol at 70°C, 2:1 molar ratio of the organic carbonate to glycerol and acetonitrile as solvent.

5.5.2 *Glycerol Carbonate Through Reaction with Urea*

Heating urea with glycerol produces GC and ammonia (Scheme 5.21). The reaction is not thermodynamically favoured at room temperature and atmospheric pressure, but the equilibrium constant increases dramatically with the rising of the reaction temperature and reduction of the pressure. In addition, removal of the ammonia

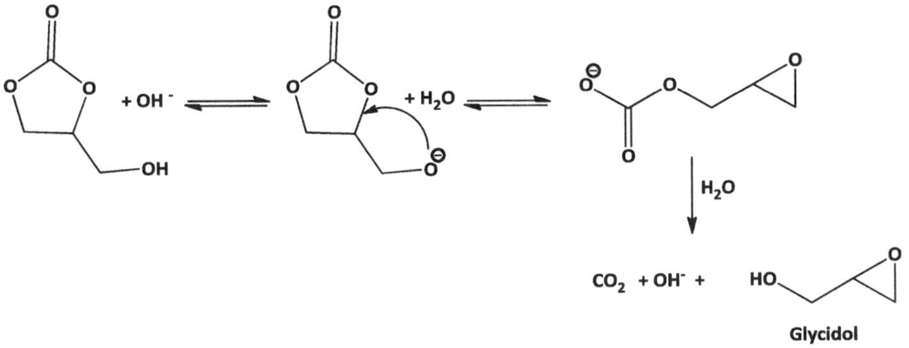

Scheme 5.20 Base-catalysed decomposition of glycerol with formation of glycidol

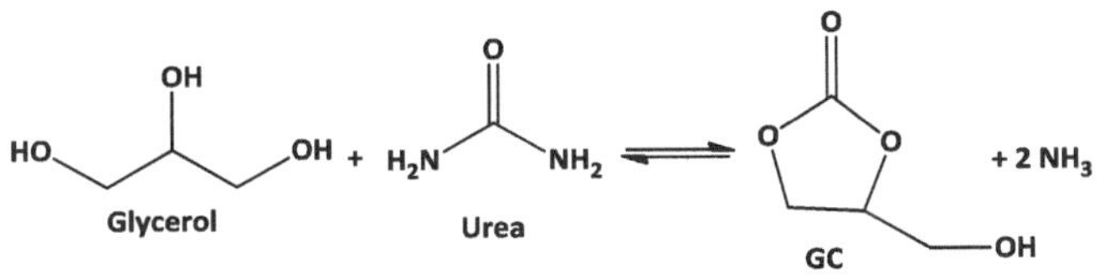

$\Delta G = +7.7$ kcal/mol

Scheme 5.21 Reaction of glycerol with urea to produce glycerol carbonate

formed may also contribute to achieve high yields of GC. Zn, Sn, La and other metal salts are normally used as weak Lewis acid catalysts.

Heterogeneous catalysts have been already reported to catalyse the reaction. The best results were found with gold impregnated on zeolite ZSM-5, which gave around 40% yield of GC [66]. The use of heterogeneous catalytic system allows the possibility of recycling, without expressive loss of the material.

A variant to the synthesis of GC with urea involves the reaction of glycerol with carbonyldiimidazole (CDI) (Scheme 5.22). The reaction proceeds fast at room temperature without any catalyst, producing quantitative yields of GC within 15 min. Crude glycerine of biodiesel production can be used without any significant drop in the conversion. The main inconvenient is the difficult separation of the imidazole from the glycerol carbonate, as well as the high costs of CDI [67].

The higher reactivity of CDI compared with urea relies on the aromatic nature of the imidazole ring. Hence, the nonbonded electron pair of the nitrogen atom linked to the carbonyl group is part of the electronic system of the aromatic ring and does not undergo resonance with the carbonyl group, as it does in urea.

5.5.3 Glycerol Carbonate Through the Direct Carbonation of Glycerol with CO_2

Although the direct carbonation of glycerol with CO_2 is not much favourable by thermodynamics, it has the great attractiveness of the direct consumption of CO_2. Dimethyl carbonate and urea can be produced from CO_2, but the processes are high-energy demanding, which ultimately neutralize the carbon balance.

Scheme 5.22 Production of glycerol carbonate from CDI and glycerol

Table 5.5 Yield of glycerol carbonate over different catalysts

Catalyst	Yield of GC (%)
NaY	0.0
AgY[a]	5.6
AgY[b]	3.0
ZnY[a]	5.8
SnY[a]	5.1
$AgNO_3$[c]	3.5
$Zn(NO_3)_2$[c]	2.7
$SnCl_2$[c]	2.6

[a]Prepared by wet impregnation followed by calcination at 450°C for 3 h
[b]Prepared by ion exchange followed by calcination at 450°C for 3 h
[c]Calcined at 450°C for 3 h

The carbonation of glycerol with CO_2 has been studied in the presence of alkyltin methoxides as catalysts [68, 69]. The maximum yield of 5.5% was achieved after 15 h of reaction time. This value is close to the thermodynamic equilibrium conversion at the temperature and pressure used.

Metal-impregnated zeolite catalysts have been tested as heterogeneous catalysts for the direct carbonation of glycerol. The catalysts were prepared through the wet impregnation of zeolite NaY with soluble metal salts, followed by calcination. Reactions were carried out in a closed vessel, at 170°C, 100 bar and 3 h. Table 5.5 shows the yield of GC for all catalytic system used. The parent NaY was not active, whereas the calcined salt precursors showed some activity, probably due to their transformation into the respective metal oxides. The metal-impregnated zeolites yielded GC at practically the equilibrium conversion, but within 3 h of reaction time, with the exception of silver-exchanged zeolite. Figure 5.9 shows the results in terms of turnover number, normalizing the GC yield by the amount of metal present in the reaction system. It can be seen that the metal-impregnated zeolites are significantly more active than the calcined salts. Further analysis revealed that the high activity of the metal-impregnated zeolites is related with the formation of dispersed metal oxide phase at the outer surface of the zeolite [70].

The mechanism of the reaction is not completely clear, but seems to involve the formation of glycerol alkoxides on the catalyst surface, which are then carbonated with CO_2. A further nucleophilic attack of the carbonyl by one of the hydroxyl groups of glycerol, followed by elimination, affords glycerol carbonate (Scheme 5.23).

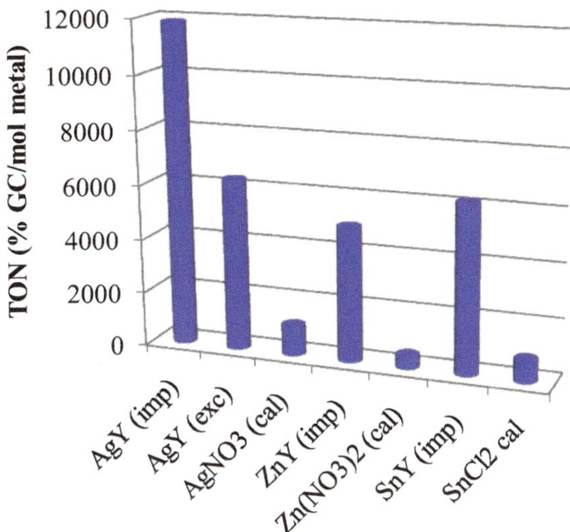

Fig. 5.9 Turnover number (TON), expressed as yield of GC (%) per mol of the metal on the reaction system

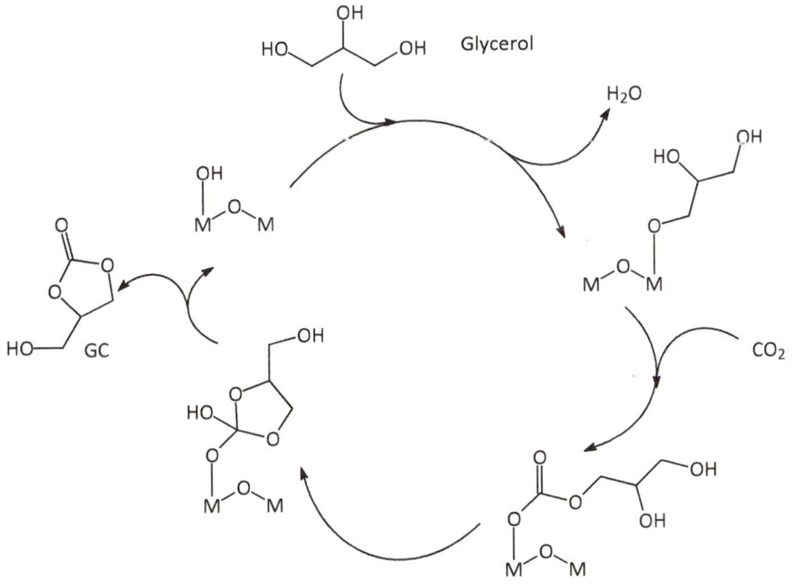

Scheme 5.23 Possible mechanistic pathway of glycerol carbonate formation on metal oxide catalysts (Reprinted from Peçanha L O, Piantonzzi R, Miranda J L Machado L C, Turci C C, Guerra A C O, Souza-Aguiar E F, Mota C J A (2015) Metal-impregnated zeolite Y as efficient Catalyst for the direct carbonation of glycerol with CO_2. Appl Catal A 504: 187–191, 71 with permission of Elsevier)

5.6 Glycerol Oxidation

5.6.1 Selective Oxidation of Glycerol

The oxidation of glycerol generates an extensive variety of products, all with high-added value. The reactions normally occur in aqueous solution, due to the high viscosity of glycerol, in the presence of molecular oxygen and temperatures ranging from 50°C to 100°C. Pd and Pt metal catalysts are among the most active for the oxidation of alcohols and polyols, since other metals undergo fast deactivation. In general, the conversion and product distribution depend on the nature of the catalyst, the reaction conditions and the oxidant source [71].

The catalytic oxidation of aqueous solutions of glycerol using cheap and clean oxidizing agents, such as air or oxygen, and recyclable catalysts provides the low-cost, environmentally friendly route to valuable specialty chemicals. Scheme 5.24 shows the diversity of products that can be obtained from glycerol oxidation. Some of them can be converted into others, and there are several ways to obtain the same product. Therefore, it is necessary to develop active but highly selective catalysts that could afford specific products [32].

The oxidation of the primary hydroxy group of glycerol leads to the formation of glyceric acid (III) and tartronic acid (IV). They can be obtained through the oxidation of glycerol in the presence of nitric acid. In addition, glyceraldehyde (II) is an intermediate in the oxidation of glycerol to glyceric acid. Oxidation of the secondary hydroxyl produces dihydroxyacetone (V), which can be further oxidized to hydroxypyruvic (VI) and mesoxalic (VII) acids. Oxalic (IX) and glycolic acids (VIII) are formed through the cleavage of C-C bonds under highly oxidant conditions [72].

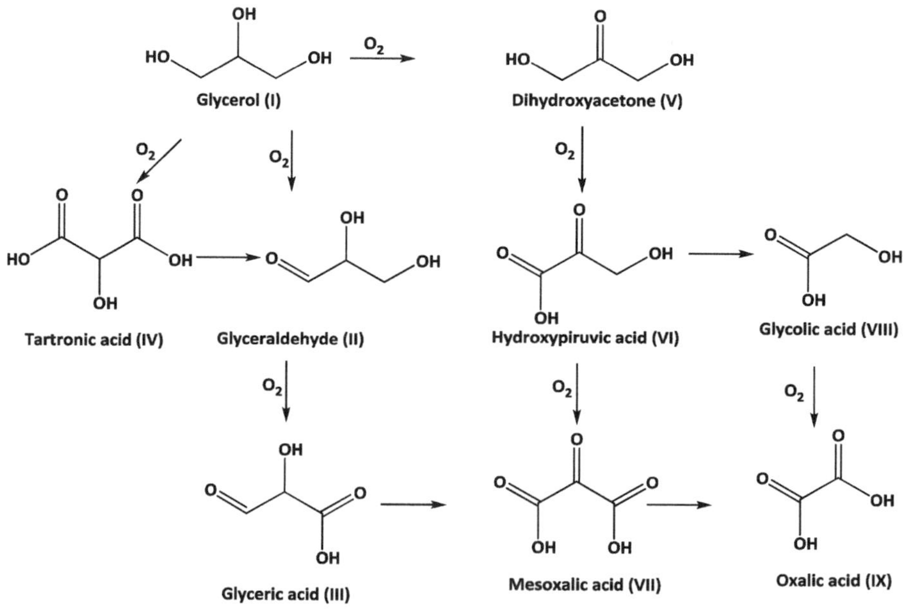

Scheme 5.24 Possible products obtained from the catalytic oxidation of glycerol

The products of glycerol oxidation find many applications, especially in pharmaceutical, cosmetics and personal care product formulations. A brief description of some applications of these products is listed:

Glyceric acid (2,3-dihydroxypropionic acid) is a natural organic acid that exists in some plants as a minor constituent. It can be converted into biodegradable polymers and emulsifiers. In addition, glyceric acid can be used in the treatment of skin problems [73].

Tartronic acid (2-hydroxypropanedioic acid) is widely used as a pharmaceutical and anticorrosive protective agent, as well as in the production of biopolymers. It is an important commercial fine chemical [74].

Dihydroxyacetone (1,3-dihydroxypropan-2-one) is a raw material in the production of self-tanning creams and lotions, in which it is present in concentrations of 3–5%. In addition to its use in the cosmetic sector, this compound is also used as a monomer for the production of polymeric biomaterials and as building block in the synthesis of other fine chemicals.

It is expected that the demand for dihydroxyacetone will increase in the coming years, which makes this chemical highly interesting to the chemical industry, because its price is around 250–500 times higher than the price of glycerol itself. Thus, any process to selectively transform glycerol into dihydroxyacetone can make a positive economic impact [75].

Mesoxalic acid (oxopropanedioic acid) is an important ligand in organic synthesis, due to its high functionality. Recently, it has shown activity as an anti-HIV agent.

Hydroxypyruvic acid (3-hydroxy-2-oxopropanoic acid) is a flavour component in cheese and has been used for synchronization of fruit maturation [76].

Table 5.6 lists some typical results of catalytic oxidation of glycerol reported in the literature.

Table 5.6 Some typical results of catalytic oxidation of glycerol

Catalysts	Conditions	Conversion (%)	Selectivity[a] (%)	Reference
5% Pd/C	60°C, 10% glycerol, 30% NaOH, 5 h, air (0.1 Mpa), pH 11	100	8 (DHA)	[77]
Bi-Pt/C	60°C, 10% glycerol, 5 h, air (0.1 Mpa), pH 2	75	50 (DHA)	[77]
5% Pt/C	0.3 M glycerol, 80°C, 3 atm of O_2, metal/ glycerol = 500 mol/mol, 4 h	83	68 (GA + TA)	[77]
1% Au/ graphite	12 mmol glycerol, 60°C, 3 h, 1500 rpm, 220 mg catalyst, O_2 (3 bar), pH 12	54	100 (GA)	[78]
Pt/ C	12 mmol glycerol, 60°C, 3 h, 1500 rpm, 220 mg catalyst, O_2 (3 bar), pH 12	63	74 (GA); 21 (GC)	[78]
5 wt.% Pd/C	50 mmol glycerol, 60°C, 1 bar, 30 mL min^{-1}, 21 h, pH 11	48	32 (GA)	[79]

(continued)

Table 5.6 (continued)

Catalysts	Conditions	Conversion (%)	Selectivity[a] (%)	Reference
5 wt.% Pd/C	50 mmol glycerol, 60°C, 1 bar, 100 mL min^{-1}, 8 h, pH 11	54	4 (GA)	[79]
5 wt.% Pt/C	50 mmol glycerol, 60°C, 1 bar, 30 mL min^{-1}, 8 h, pH 11	60	48 (GA)	[79]
5 wt.% Pt/C	50 mmol glycerol, 60°C, 1 bar, 100 mL min^{-1}, 8 h, pH 11	89	41 (GA)	[79]
5 wt.% Au/graphite	6 mmol glycerol, 60°C, 6 bar, 100 mL min^{-1}, 8 h, pH 12	91	92 (GA)	[79]
Pt/MCN[b]	0.3 M glycerol, 60°C, 0.3 Mpa O_2, 4 h	69	36 (GA); 18 (DHA)	[80]
Au/HSA[c]-CeO_2	1.5 M glycerol, 60°C, 1 atm O_2, 5 h	40	45 GA; 25 (DHA)	[81]
4.8 Pt/Al_2O_3	1 atm O_2, 60°C, 7 h, without base addition	49	47 (GA); 8 (DHA)	[82]
4.8 Pt/SiO_2	1 atm O_2, 60°C, 7 h, without base addition	21	28 (GA); 12 (DHA)	[82]
4.8 Pt/C	1 atm O_2, 60°C, 7 h, without base addition	22	29 (GA); 10 (DHA)	[82]

[a]*DHA* dihydroxyacetone, *GA* glyceric acid, *TA* tartronic acid, *GC* glyceraldehyde
[b]*MCN* mesoporous carbon nitride
[c]*HAS* high specific surface areas

References

1. Romagnoli P (2007) Fine chemicals: the industry and the business. J Generic Med 5:91
2. Chandrasekhar S (1987) Product stability in kinetically-controlled organic reactions. Chem Soc Rev 16:313–338
3. Deutsch J, Martin A, Lieske H (2007) Investigations on heterogeneously catalyzed condensations of glycerol to cyclic acetals. J Catal 245:428–435
4. da Silva CXA, Gonçalves VLC, Mota CJA (2009) Water-tolerant zeolite catalyst for the acetalisation of glycerol. Green Chem 11:38–41
5. Smit B, Maessen TLM (2008) Towards a molecular understanding of shape selectivity. Nature 451:671–678
6. Okuhara T (2002) Water-tolerant solid acid catalysts. Chem Rev 102:3641–3666
7. Silva PH, Gonçalves VLC, Mota CJA (2010) Glycerol acetals as anti-freezing additives for biodiesel. Bioresour Technol 101:6225–6229
8. Dodson JR, Avellar T, Athayde J, Mota CJA (2014) Glycerol acetals with antioxidant properties. Pure Appl Chem 86:905–912
9. Piasecki A, Sokolowski A, Burczyk B, Kotlewska U (1997) Synthesis and surface properties of chemodegradable anionic surfactants: sodium (2-n-Alkyl-1,3-Dioxan-5-yl) sulfates. J Am Oil Chem Soc 74:33
10. Mota CJA, da Silva CXA, Rosenbach N Jr, Costa J, da Silva F (2010) Glycerin derivatives as fuel additives: the addition of glycerol/acetone ketal (solketal) in gasolines. Energy Fuel 24:2733–2736
11. Ahmed FE (2001) Toxicology and human health effects following exposure to oxygenated or reformulated gasoline. Toxicol Lett 123:89–113

12. Ozorio LP, Pianzolli R, Mota MBS, Mota CJA (2012) Reactivity of glycerol/acetone ketal (solketal) and glycerol/formaldehyde acetals toward acid-catalyzed hydrolysis. J Braz Chem Soc 23:931–937
13. da Silva CXA, Mota CJA (2011) The influence of impurities on the acid-catalyzed reaction of glycerol with acetone. Biomass Bioenergy 35:3547–3551
14. Melero JA, Vicente G, Morales G, Paniagua M, Bustamante J (2010) Oxygenated compounds derived from glycerol for biodiesel formulation: influence on EN 14214 quality parameters. Fuel 89:2011–2018
15. Vicente G, Melero JA, Morales G, Paniagua M, Martín E (2010) Acetalisation of bio-glycerol with acetone to produce solketal over sulfonic mesostructured silicas. Green Chem 12:899–907
16. Nanda MR, Yuan Z, Qin W, Ghaziaskar HS, Poirier MR, Xu CC (2014) Catalytic conversion of glycerol to oxygenated fuel additive in a continuous flow reactor: process optimization. Fuel 128:113–119
17. Manjunathan P, Maradur SP, Halgeri AB, Shanbhag GV (2015) Room temperature synthesis of solketal from acetalization of glycerol with acetone: effect of crystallite size and the role of acidity of beta zeolite. J Mol Catal A Chem 396:47–54
18. Li L, Korányi TI, Sels BF, Pescarmona PP (2012) Highly-efficient conversion of glycerol to solketal over heterogeneous Lewis acid catalysts. Green Chem 14:1611
19. Khayoon MS, Hameed BH (2013) Solventless acetalization of glycerol with acetone to fuel oxygenates over Ni–Zr supported on mesoporous activated carbon catalyst. Appl Catal A Gen 464–465:191–199
20. Umbarkar SB, Kotbagi TV, Biradar AV, Pasricha R, Chanale J, Dongare JK, Mamede NS, Lancelot C, Payen E (2009) Acetalization of glycerol using mesoporous MoO_3/SiO_2 solid acid catalyst. J Mol Catal A Chem 310:150–158
21. Sudarsanam P, Mallesham B, Prasad AN, Reddy PS, Reddy BM (2013) Synthesis of bio-additive fuels from acetalization of glycerol with benzaldehyde over molybdenum promoted green solid acid catalysts. Fuel Process Technol 106:539–545
22. Gadamsetti S, Rajan NP, Rao GS, Char KVR (2015) Acetalization of glycerol with acetone to bio fuel additives over supported molybdenum phosphate catalysts. J Mol Catal A Chem 410:49–57
23. Reddy PS, Sudarsanam P, Mallesham B, Raju G, Reddy BM (2011) Acetalisation of glycerol with acetone over zirconia and promoted zirconia catalysts under mild reaction conditions. J Ind Eng Chem 17:377–381
24. Kiatkittipong W, Suwanmanee S, Laosiripojana N, Praserthdam P, Assabumrungrat S (2010) Cleaner gasoline production by using glycerol as fuel extender. Fuel Process Technol 91:456–460
25. Mota CJA, Gonçalves VLC (2007) Process of glycerol etherification and fuel additives. PI 0700063-4 (Brazil)
26. Pariente S, Tanchoux N, Fajula F (2009) Etherification of glycerol with ethanol over solid acid catalysts. Green Chem 11:1256–1261
27. Pico MP, Romero A, Rodríguez S, Santos A (2012) Etherification of glycerol by tert-butyl alcohol: kinetic model. Ind Eng Chem Res 51:9500–9509
28. Yuan Z, Xia S, Chen P, Hou Z, Zheng X (2011) Etherification of biodiesel-based glycerol with bioethanol over tungstophosphoric acid to synthesize glyceryl ethers. Energy Fuel 25:3186–3191
29. Melero JA, Vicente G, Paniagua M, Morales G, Muñoz P (2012) Synthesis of oxygenated compounds for fuel formulation: etherification of glycerol with ethanol over sulfonic modified catalysts. Bioresour Technol 103:142–151
30. Kesling HS, Karas LJ, Liotta FJ (2014) Diesel fuel. US patent 5308365
31. Noureddini H (2000) Process for producing biodiesel fuel with reduced viscosity and a cloud point below thirty-two (32) degrees Fahrenheit. US patent 6015440
32. Zheng Y, Chen X, Shen Y (2008) Commodity chemicals derived from glycerol, an important biorefinery feedstock. Chem Rev 108:5253–5277

33. Pinto BP, Lyra JT, Nascimento JAC, Mota CJA (2016) Ethers of glycerol and ethanol as bioadditives for biodiesel. Fuel 68:76–80
34. Martin A, Richter M (2011) Oligomerization of glycerol–a critical review. Eur J Lipid Sci Technol 113:100–117
35. Soi HS, Bakar ZA, Din NS, Idris Z, Kian YS, Hassan HA, Ahmad S (2014) Process of producing polyglycerol from crude glycerol. US Patent 20110190545 A1
36. Ayoub M, Abdullah AZ (2013) Diglycerol synthesis via solvent-free selective glycerol etherification process over lithium-modified clay catalyst. Chem Eng J 225:784–789
37. García-Sancho C, Moreno-Tost R, Mérida-Robles JM, Santamaría-González J, Jiménez-López A, Torres PM (2011) Etherification of glycerol to polyglycerols over MgAl mixed oxides. Catal Today 167:84–90
38. Corma A, Iborra S, Velty A (2007) Chemical routes for the transformation of biomass into chemicals. Chem Rev 107:2411
39. Barrault J, Pouilloux Y, Clacens JM, Vanhove C, Bancquart S (2002) Catalysis and fine chemistry. Catal Today 75:177–181
40. Clacens JM, Pouilloux Y, Barrault J (2002) Selective etherification of glycerol to polyglycerols over impregnated basic MCM-41 type mesoporous catalysts. Appl Catal A Gen 227:181
41. Clacens JM, Pouilloux Y, Barrault J (2002) Synthesis and modification of basic mesoporous materials for the selective etherification of glycerol. Stud Surf Sci Catal 143:687
42. Richter M, Krisnandi YK, Eckelt R, Martin A (2008) Homogeneously catalyzed batch reactor glycerol etherification by $CsHCO_3$. Catal Commun 9:2112–2116
43. Ruppert AM, Meeldijk JD, Kuipers BWM, Ern BH, Weckhuysen BM (2016) Glycerol etherification over highly active CaO-based materials: new mechanistic aspects and related colloidal particle formation. Chem Eur J 14:2024
44. García E, Laca M, Pérez E, Garrido A, Peinado J (2008) New class of acetal derived from glycerin as a biodiesel fuel component. Energy Fuel 22:4274–4280
45. Galan MI, Bonet J, Sire R, Reneaume JM, Plesu AE (2009) From residual to useful oil: revalorization of glycerine from the biodiesel synthesis. Bioresour Technol 100:3775–3778
46. Gonçalves VLC, Pinto BP, da Silva JFC, Mota CJA (2008) Acetylation of glycerol catalyzed by different solid acids. Catal Today 133–135:673–677
47. Silva LN, Gonçalves VLC, Mota CJA (2010) Catalytic acetylation of glycerol with acetic anhydride. Catal Commun 11:1036–1039
48. Bender ML, Ladenheim H, Chen MC (1961) Acylium ion formation in the reactions of carboxylic acid derivatives. II. The hydrolysis and oxygen exchange of methyl mesitoate in sulfuric acid. J Am Chem Soc 83:123–127
49. Balaraju M, Nikhitha P, Jagadeeswaraiah K, Srilatha K, Prasad PSS, Lingaiah N (2010) Acetylation of glycerol to synthesize bioadditives over niobic acid supported tungstophosphoric acid catalysts. Fuel Process Technol 91:249–253
50. Sánchez JA, Hernández DL, Moreno JA, Mondragón F, Fernández JJ (2011) Alternative carbon based acid catalyst for selective esterification of glycerol to acetylglycerols. Appl Catal A Gen 405:55–60
51. Liao X, Zhu Y, Wang SG, Li Y (2010) Theoretical elucidation of acetylating glycerol with acetic acid and acetic anhydride. Appl Catal B 94:64–70
52. Rezayat M, Ghaziaskar HS (2009) Continuous synthesis of glycerol acetates in supercritical carbon dioxide using amberlyst-15. Green Chem 11:710–715
53. Wang ZQ, Zhang Z, Wu WJ, Li LD, Zhang MH, Zhang ZB (2016) A swelling-changeful catalyst for glycerol acetylation with controlled acid concentration. Fuel Process Technol 142:228–234
54. Liao X, Zhu Y, Wang S-G, Li Y (2009) Producing triacetylglycerol with glycerol by two steps: esterification and acetylation. Fuel Process Technol 90:988–993
55. Testa ML, Parola VL, Liotta LF, Venezia AL (2013) Screening of different solid acid catalysts for glycerol acetylation. J Mol Catal A Chem 367:69–76
56. Konwar LJ, Arvela PM, Begum P, Kumar N, Thakur AJ, Mikkola JP, Deka RC, Deka D (2015) Shape selectivity and acidity effects in glycerol acetylation with acetic anhydride: selective synthesis of triacetin over Y-zeolite and sulfonated mesoporous carbons. J Catal 329:237–247

57. Patel A, Singh S (2014) A green and sustainable approach for esterification of glycerol using 12-tungstophosphoric acid anchored to different supports: kinetics and effect of support. Fuel 118:358–364
58. Sandesh S, Manjunathan P, Halgeri AB, Shanbhag GV (2015) Glycerol acetins: fuel additive synthesis by acetylation and esterification of glycerol using cesium phosphotungstate catalyst. RSC Adv 5:104354–104362
59. Sun J, Tong T, Yu L, Wan J (2016) An efficient and sustainable production of triacetin from the acetylation of glycerol using magnetic solid acid catalysts under mild conditions. Catal Today 264:115–122
60. Pagliaro M, Ciriminna R, Kimura H, Rossi M, Pina CD (2009) Recent advances in the conversion of bioglycerol into value-added products. Eur J Lipid Sci Technol 111:788–799
61. Melero JA, Grieken R, Morales R, Paniagua M (2007) Acidic mesoporous silica for the acetylation of glycerol: synthesis of bioadditives to petrol fuel. Energy Fuel 21:1782–1791
62. Dodson J, Leite TMC, Pontes NS, Pinto BP, Mota CJA (2014) Green acetylation of solketal and glycerol formal by heterogeneous acid catalysts to form a biodiesel fuel additive. ChemSusChem 7:2728–2734
63. Li J, Wang T (2011) Chemical equilibrium of glycerol carbonate synthesis from glycerol. J Chem Thermodyn 43:731–736
64. Teng WK, Ngoh GC, Yusoff R, Aroua MK (2014) A review on the performance of glycerol carbonate production via catalytic transesterification: effects of influencing parameters. Energy Convers Manag 88:484–497
65. Jung H, Lee Y, Kim D, Han SO, Kim SW, Lee J (2012) Enzymatic production of glycerol carbonate from by-product after biodiesel manufacturing process. Enzym Microb Technol 51:143–147
66. Hammond C, Lopez-Sanchez JA, Rahim MHA, Dimitratos N, Jenkins RL, Carley AF, He Q, Kiely CJ, Knight DW, Hutchings GJ (2011) Synthesis of glycerol carbonate from glycerol and urea with gold-based catalysts. Dalton Trans 40:3927–3937
67. Mota CJA, Gonçalves VLC, Rodrigues RC (2007) Process of production of glycerol carbonate. PI 0706121-8 (Brazil)
68. Aresta M, Dibenedetto A, Pastore C (2006) Direct carboxylation of alcohols to organic carbonates: comparison of the Group 5 element alkoxides catalytic activity: an insight into the reaction mechanism and its key steps. Catal Today 115:88–94
69. Aresta M, Dibenedetto A, Nocito F, Pastore C (2006) A study on the carboxylation of glycerol to glycerol carbonate with carbon dioxide: the role of the catalyst, solvent and reaction conditions. J Mol Catal A 257:149–153
70. Peçanha LO, Piantonzzi R, Miranda JL, Machado LC, Turci CC, Guerra ACO, Souza-Aguiar EF, Mota CJA (2015) Metal-impregnated zeolite Y as efficient catalyst for the direct carbonation of glycerol with CO_2. Appl Catal A 504:187–191
71. Hamid SBA, Basiron N, Yehye WA, Sudarsanam P, Bhargava SK (2016) Nanoscale Pd-based catalysts for selective oxidation of glycerol with molecular oxygen: structure–activity correlations. Polyhedron 120:124–133
72. Bianchi CL, Canton P, Dimitratos N, Porta F, Prati L (2005) Selective oxidation of glycerol with oxygen using mono and bimetallic catalysts based on Au, Pd, and Pt metals. Catal Today 102:203–212
73. Sato S, Kitamoto D, Habe H (2014) Chemical mutagenesis of Gluconobacter frateurii to construct methanol-resistant mutants showing glyceric acid production from methanol-containing glycerol. J Biosci Bioeng 117:197–199
74. Cai J, Ma H, Zhang J, Du Z, Huang Y, Gao J, Xu J (2014) Catalytic oxidation of glycerol to tartronic acid over Au/HY catalyst under mild conditions. Chin J Catal 35:1653–1660
75. Martínez-Gallegos JF, Burgos-Cara A, Caparrós-Salvador F, Luzón-González G, Fernández-Serrano M (2015) Dihydroxyacetone crystallization: process, environmental, health and safety criteria application for solvent selection. Chem Eng Sci 134:36–43
76. Grybek R, Johnston FB (1975) Composition and method for treating plants and trees. US Patent 3, 884: 674

77. Garcia R, Besson M, Gallezot P (1995) Chemoselective catalytic oxidation of glycerol with air on platinum metals. Appl Catal A 127:165–176
78. Carrettin S, McMorn P, Johnston P, Griffinb K, Hutchings GJ (2002) Selective oxidation of glycerol to glyceric acid using a gold catalyst in aqueous sodium hydroxide. Chem Commun 7:696–697
79. Carrettin S, McMorn P, Johnston P, Griffin K, Kiely CJ, Hutchings GJ (2003) Oxidation of glycerol using supported Pt, Pd and Au catalysts. Phys Chem Chem Phys 5:1329–1336
80. Wang FF, Shao S, Liu CL, Xu CL, Yang RZ, Dong WS (2015) Selective oxidation of glycerol over Pt supported on mesoporous carbon nitride in base-free aqueous solution. Chem Eng J 264:336–343
81. Demirel S, Kern P, Lucas M, Claus P (2007) Oxidation of mono- and polyalcohols with gold: comparison of carbon and ceria supported catalysts. Catal Today 122:292–300
82. Chornaja S, Sproge E, Dubencovs K, Kulikova L, Serga V, Cvetkovs A, Kampars V (2014) Selective oxidation of glycerol to glyceraldehyde over novel monometallic platinum catalysts. Key Eng Mater 604:138–141

Chapter 6
Glycerol Conversion in the Biorefinery Context

Abstract This last chapter discusses the opportunities and challenges of converting glycerol into different chemicals, within the biorefinery context. As most of the biodiesel plants in the world are of low to medium capacity, the surplus of glycerol is somewhat spread and available in small quantities in some regions. Therefore, logistic problems, related to the transportation, storage and purification of the glycerol must be taken into account to choose the best opportunity of investment. Specialty and fine chemicals can overcome these problems, because of the high-added value of the final products. Glycerol derivatives that are used as fuel additives may be the best choice for biodiesel producers, as they can be used in the biofuel to improve its properties. In addition, the process conditions and installations are similar to those used in the manufacture of the biodiesel itself. Finally, the use of glycerol as raw material for the production of commodities may be more suitable to traditional chemical companies. They are usually integrated companies, producing the monomers and the polymers to be used in the automotive, textile, painting and other large sectors. In addition, these companies are used to work with processes in continuous flow at high temperatures and pressures.

Keywords Glycerol • Biorefinery • Commodities • Specialty • Fine chemicals

6.1 Biorefinery Context

A biorefinery may be described as an integrated industrial plant that converts biomass or biomass-derived products into fuels, energy and chemicals. Different from an oil refinery, where the processes and stages are well established, the biorefinery processes and raw materials can be highly diversified; in addition, the plant may involve a single or a small number of units to convert the biomass.

Figure 6.1 shows a simplified scheme of an oil refinery. There are basically three stages: physical separation, chemical conversion processes and upgrading units. The first stage usually involves atmospheric and vacuum distillation, separating the major hydrocarbon fractions. The second stage normally involves the catalytic cracking or sometimes hydrocracking of the heavy fractions, especially gas oil, to mostly produce lighter derivatives, such as LPG and gasoline. Finally, in the third stage, there may occur the hydrotreatment (HDT) of some fractions, such as

C.J.A. Mota et al., *Glycerol*, DOI 10.1007/978-3-319-59375-3_6

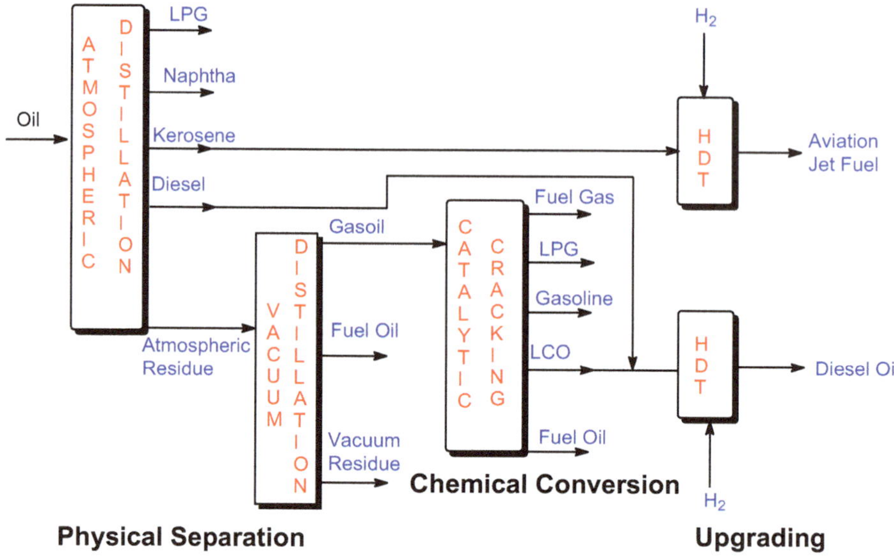

Fig. 6.1 Simplified scheme of oil refining, showing the three stages: physical separation, chemical conversion and upgrading of derivatives. LCO stands for light cycle oil, which has approximately the same boiling point range of diesel. HDT stands for hydrotreatment. LPG stands for liquefied petroleum gas, which mainly comprises propane and butane

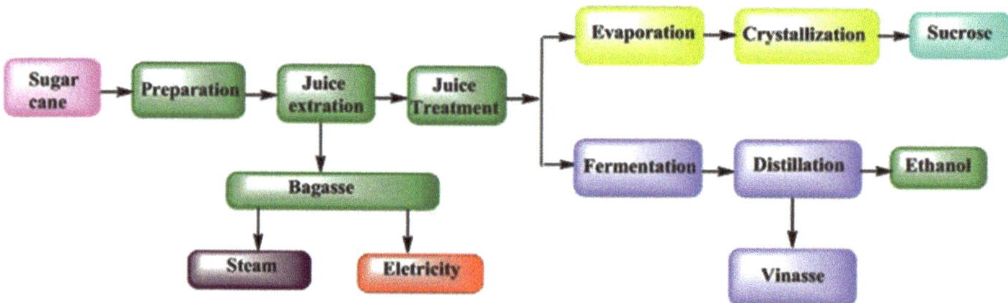

Fig. 6.2 Simplified scheme of a sugar cane-based biorefinery to produce ethanol and sucrose

kerosene and diesel, to reduce the sulphur and nitrogen content to meet required specifications.

The nature of biorefineries is different because of the plurality of the biomass processed and variety of product that could be obtained. Hence, a biorefinery based on carbohydrates, aiming to produce ethanol, may involve completely different steps and processes than a facility used to convert vegetable oils into biodiesel. Ethanol production usually involves biotechnological processes, whereas the fabrication of biodiesel is basically a thermochemical reaction. Figure 6.2 shows a simplified scheme of ethanol production from sugar cane. Initially, the sugar cane is

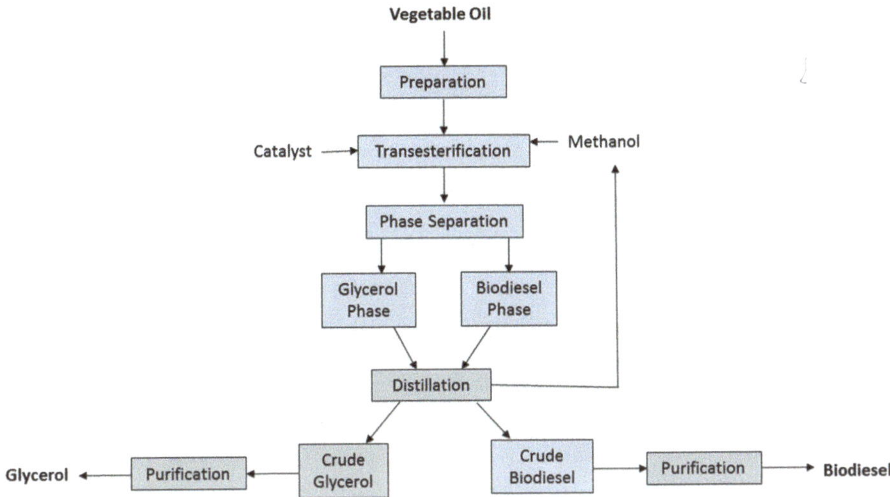

Fig. 6.3 Simplified scheme of biodiesel production through the transesterification of vegetable oils

prepared, and the juice is extracted yielding bagasse as by-product. The juice can follow two routes: it can be evaporated to crystallize the sucrose that is commercialized as common sugar, or it can be fermented to yield ethanol and vinasse, the residue of distillation. As a matter of fact, the sugar cane-based biorefinery to produce ethanol and sucrose leaves virtually no residue. The bagasse is burned to produce steam, which is used in the plant, and to generate electricity, whereas the vinasse is reutilized as a fertilizer in the soil or to generate biogas (methane) through anaerobic fermentation. The lignocellulose material of the bagasse can also be hydrolysed and fermented to produce more ethanol. This is the base of the technologies of second-generation ethanol, which is being implemented by some companies in Brazil and other countries.

The simplified scheme of biodiesel production through the transesterification of vegetable oils is presented in Fig. 6.3. Initially, the oil may be treated, for instance, to reduce acidity. Then, transesterification takes place upon heating the oil with excess of methanol and in the presence of a basic catalyst. As reaction proceeds, phase separation occurs; the upper phase corresponds to biodiesel and methanol, whereas the bottom phase is rich in glycerol but also contains methanol and the basic catalyst. After separation, the two phases are independently distilled to recover the methanol, which is usually in excess to help shifting the equilibrium of transesterification. The oil and glycerine phases are purified, yielding the biodiesel and pure glycerol for commercialization.

Biorefinery plants are usually smaller and scattered distributed compared with oil refining plants. Figure 6.4 shows a map with the location of biodiesel plants in the USA. It is clear the large number of facilities spread all over the country. The situation is similar in Brazil, which counts with around 50 biodiesel plants in

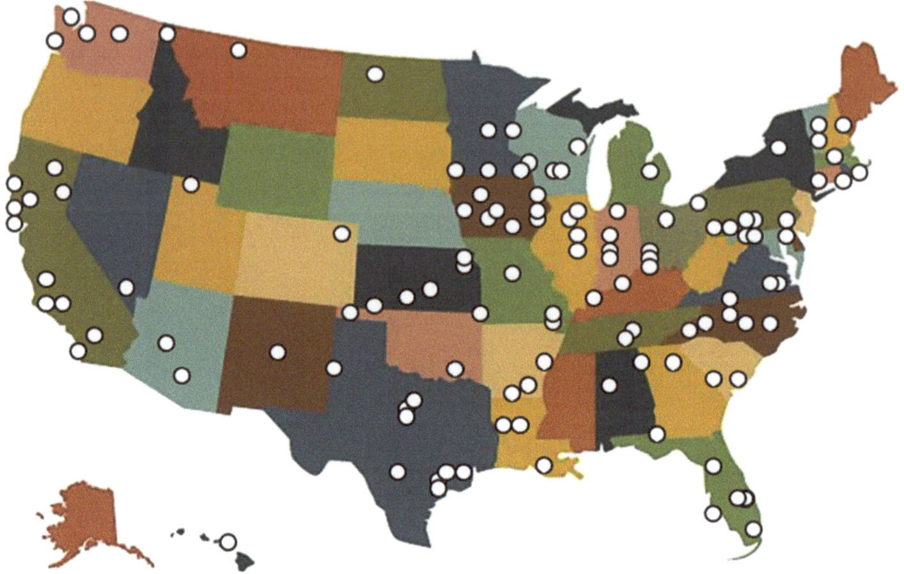

Fig. 6.4 Location of the biodiesel plants in the USA (Reprinted with permission of the Bureau of Labor Statistics and the National Biodiesel Board)

different locations of the country. Thus, any logistic concerning transportation, purification and conversion of the glycerol must be taken into account, which may impact the production costs.

6.2 Specialty Chemicals from Glycerol in a Biorefinery Context

The conversion of glycerol into ethers, acetals/ketals and esters is usually carried out in batch conditions and temperatures ranging from ambient to 200 °C. In some cases, autogenous pressure is also applied. These glycerol derivatives find applications as fuel additives, improving properties of the biodiesel, diesel oil and gasoline.

It has been shown that glycerol ethers are good additives for biodiesel [1, 2]. The reaction between glycerol and ethanol can be conveniently carried out in the same places where biodiesel is produced. The synthesized ethers can be blended with biodiesel to improve its cold flow properties. The product is completely renewable, because ethanol is obtained from fermentation of carbohydrates. Notwithstanding, ethanol is not usually employed in the transesterification of vegetable oils and fats to produce biodiesel. There are some technical difficulties, but costs are the major issue, as ethanol is still more expansive than methanol. Etherification of glycerol with methanol has also been studied. The trimethylated ether has proven to be a good additive to gasoline [3] and could be produced from the acid-catalysed reaction between glycerol and methanol. Nevertheless, since methanol is usually obtained from natural gas, the trimethyl glyceryl ether would not be a completely renewable product and, therefore, would not have the same impact as the ethyl glyceryl ethers would do as fuel additives.

Acetals and ketals are other derivatives that could be produced in biodiesel plant facilities. The reaction can be carried out at atmospheric pressure and temperatures around 60–70°C in batch reactor, but, more recently, flow conditions has also been employed [4–6]. Solketal, the product of the reaction between glycerol and acetone, has been proved to be a good fuel additive, especially for gasoline [7]. It can also be converted in solketal acetate, which has been used as additive for biodiesel [8], being a good choice of glycerol derivative that can be produced in biodiesel plants.

Nowadays, acetone is produced from the acid decomposition of cumene hydroperoxide, being dependent on fossil sources. Nevertheless, acetone can also be produced from sugar fermentation in the so-called ABE fermentation process that also produces butanol and ethanol [9]. This pathway is no longer used in industrial high-scale production but has recently gained interest due to the potential use of butanol as a renewable fuel additive. Some pilot and small industrial scale plants have been constructed, but the technology is not fully optimized yet to be competitive with fermentation processes to produce ethanol. The interest in butanol as fuel additive relies on its higher energetic content and better compatibility with hydrocarbons compared with ethanol. Thus, any important development in the ABE technology to lower the costs of butanol production must involve the utilization of acetone, which is a by-product of this fermentative pathway.

Acetals of glycerol are less studied as fuel additives. The acetal produced from the reaction of glycerol and butanal has been shown to improve the cold flow properties of the biodiesel [10]. However, the main route for the industrial production of butanal involves the hydroformylation of propene and relies on fossil sources (Scheme 6.1). This is also the present main route to butanol, which is obtained upon the hydrogenation of butanal. If butanol could be efficiently and cheaply produced from the ABE fermentation process, then butanal could be produced from the oxidation of butanol, but this route is still not commercially available.

Other specialty products from glycerol can be of great interest. Perhaps not to the biodiesel producers but for small- or medium-size companies, focused on technology and small scale plants. Glycerol carbonate (GC) and dihydroxyacetone (DHA) are the main candidates in this category.

Glycerol carbonate is a product that is gaining increased applications each year. Today, it may find applications as special solvent, in the production of polycarbonates,

Scheme 6.1 Industrial route to the production of butanal and butanol

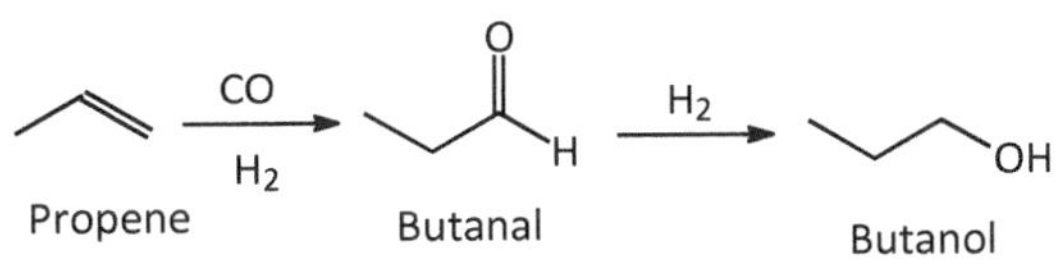

Propene $\xrightarrow[H_2]{CO}$ Butanal $\xrightarrow{H_2}$ Butanol

electrolyte in batteries and in formulations of personal care products. The major industrial route for the production of this cyclic organic carbonate is the reaction of glycerol with urea. The process is carried out at batch conditions, temperatures between 150 and 180°C and reduced pressure to remove the ammonia formed as by-product. This pathway does not lead to a completely renewable GC, because urea is produced from natural gas, involving a highly intensive energy process. Therefore, this would not be the best pathway in the context of a biorefinery.

The other route to GC is the direct carbonation of glycerol with CO_2 (Scheme 6.2). The main drawback is thermodynamic, which limits the yield of GC within 5 to 10%, depending on the reaction conditions. Therefore, technological developments to shift equilibrium are extremely necessary to the industrial feasibility of this route.

The reaction of CO_2 and glycerol is usually carried out in temperatures ranging from 140 to 200°C and pressures between 30 and 120 bar, approximately, in the presence of a proper catalyst. At these conditions, removal of the water formed by any physical process, such as distillation and adsorption, is virtually unfeasible. Hence, the most traditional way to push equilibrium towards products involves the addition of water suppressor agents, like nitriles, acetals and oxiranes [11–13]. The main drawback is the production of large amounts of undesired by-products, formed upon the hydrolysis of the suppressor agents, making the whole process not much sustainable.

A study on the use of methyl trihaloacetic esters as water suppressor agents in the direct carbonation of methanol to dimethyl carbonate (DMC) (Scheme 6.3) has been recently reported [14]. The main advantage would be the potential reutilization of the carboxylic acid formed upon the ester hydrolysis, which can be further reconverted in the ester, yielding, virtually, no by-product or waste. In addition, the methanol released upon the hydrolysis may also help shifting the equilibrium towards DMC. A similar approach can be used to the synthesis of glycerol carbonate, probably using the glycerol trihaloacetatic esters as water suppressors. However, further

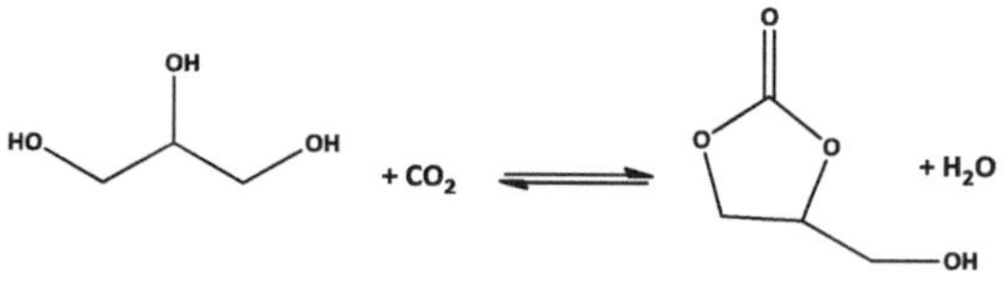

Scheme 6.2 Direct carbonation of glycerol with CO_2

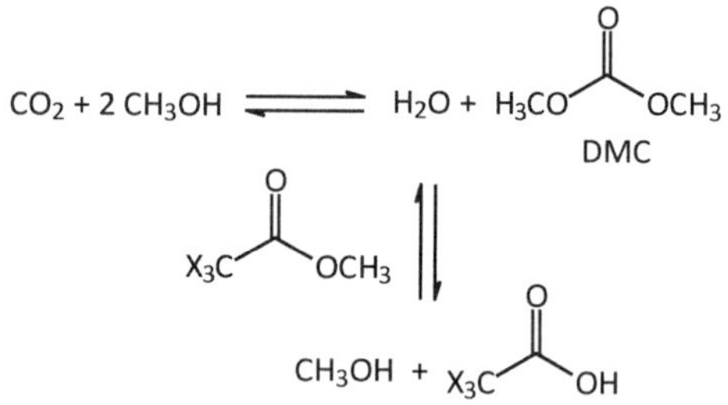

Scheme 6.3 Use of methyl trihaloacetates as water suppressor agents in the synthesis of DMC from CO_2 and methanol. A potential strategy in the synthesis of GC from CO_2 and glycerol to overcome thermodynamic limitations

$$C_6(H_2O)_6 \longrightarrow 2\ \text{CH}_3\text{CH}_2\text{OH} + 2\ CO_2$$

Scheme 6.4 Fermentation of glucose to produce ethanol and CO_2

developments must be taken in the catalytic system before a commercial process can be implemented.

Another issue is the source of the CO_2. There are many technologies for carbon dioxide capture in post-combustion systems, such as the use of solvents, adsorbents and membranes. Nevertheless, the costs of capture and purification are still high, affecting any further use of the CO_2. The costs of capture can be dramatically reduced with the implementation of precombustion systems, where the fuel is initially transformed in CO_2 and H_2 before energy generation, because of the higher concentration and pressure of the CO_2 gas stream.

Another alternative source of CO_2 is the fermentative processes. For instance, in glucose fermentation to produce ethanol, one-third of the carbon atoms are released as CO_2 (Scheme 6.4). In Brazil for instance, this CO_2 from sugar cane fermentation is not used, being mostly released to the atmosphere. The costs of capturing and converting this CO_2 are much lower than those of flue gas in post-combustion systems. In addition, the gas is in high purity and does not require many purification steps. The main drawback is the seasonality of the ethanol industry, because fermentation can only be carried out after plantation, growth and harvesting of the crop. The second-generation ethanol may change this situation, enabling the production throughout the year.

Dihydroxyacetone (DHA) is the main ingredient in self-tanning formulations, and its price may be 500-fold higher than the price of glycerol, making it highly interesting for a small business company. It may be produced either by biotechnological or thermochemical conversions of glycerol. The former route is more attractive and feasible to be implemented in an industrial scale. The thermocatalytic oxidation of glycerol still lacks selectivity, and many products can be formed, which increases the costs of separation and purification. On the other hand, the biotechnological route can be more efficient, resulting in a product with few impurities. For instance, *Gluconobacter oxydans* produces DHA with a selectivity around 75% [15–17].

6.3 Commodity Chemicals from Glycerol in a Biorefinery Context

High-scale chemical plants produce commodities that are mostly converted to polymers. Today, most of the plastics depend on fossil sources, but renewable raw materials are gaining significant importance in the production of commodities and basic chemicals. For instance, ethanol has been used to commercially produce polyethylene, upon its acid-catalysed dehydration. Costs are still a major concern, because the price of biomass-derived feedstock is usually higher compared with raw materials from fossil sources.

Propylene and propylene-based products are an important segment of the petrochemical industry. Besides polypropylene, which is produced from the polymerization of propene or propylene in the presence of a proper catalyst, other monomers, such as epichlorohydrin, acrylic acid and propylene glycol, among others, are produced in large scale from propene itself.

The production of commodity chemicals from glycerol of biodiesel production is more suitable to traditional petrochemical companies, which are normally integrated, producing the monomers and the polymers. In addition, the processes normally operate in continuous flow conditions, at high pressures and temperatures. Therefore, knowledge and optimization of process engineering are required, making traditional chemical companies more adapted to implement these technologies.

Chapter 4 highlighted some of the main commodity chemicals that can be produced from the thermocatalytic conversion of glycerol. Processes for the production of 1,2-propanediol and epichlorohydrin have been industrially implemented and will not be discussed here. They are successful examples on how glycerol can replace fossil-derived technologies in the chemical industry and bring sustainability together with economic viability.

6.3.1 Glycerol Hydrogenolysis to Propene in the Context of a Biorefinery

Propene is one of the main chemical commodities produced in the world. About 2/3 of the propene is transformed into polypropylene, a versatile polymer used in many sectors. Today, there are few technological pathways for the production of propene from renewable feedstock. Because glycerol has three carbon atoms in its structure, it looks like the preferable biomass-derived molecule to be converted in propene.

A pioneer technology of glycerol hydrogenolysis to propene over supported Fe-Mo catalysts has been developed between the former Quattor Petrochemicals (now part of Braskem conglomerate) and the Federal University of Rio de Janeiro [18–20]. The process may be operated in continuous flow conditions, using concentrated glycerol aqueous solution, temperatures ranging from 275 to 325°C and pressures up to 10 bar. At optimized conditions, complete glycerol conversion with 90% selectivity to propene can be observed. Methane, ethane and propane are formed in minor amounts, as well as oxygenated intermediates, such as propanediols, propanol and acetone. Water is the major by-product and, together with the oxygenated intermediates, must be separated from the propene stream, because the catalysts used in the polymerization process are sensitive to water and alcohols.

Figure 6.5 shows a simplified scheme of the concept of the entire process. The reaction would be carried out in a fixed bed reactor operating at low pressures (up to five or eight bar). Water and heavy oxygenated intermediates can be separated by condensation, leaving a gas stream rich in hydrogen, propene, propane, methane and ethene. The light gases would be then removed through traditional low tempera-

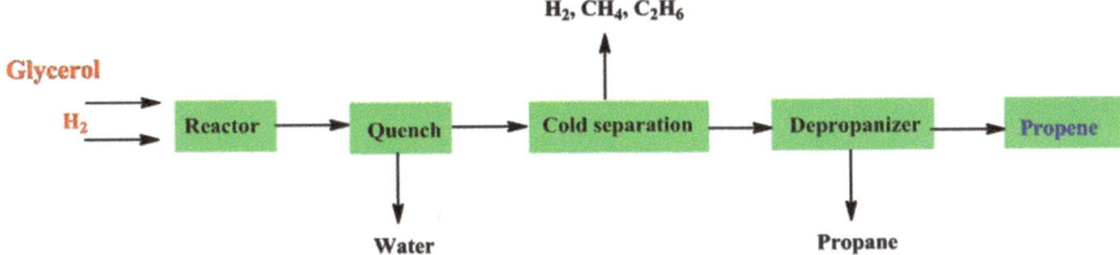

Fig. 6.5 Simplified schematic concept of the glycerol hydrogenolysis process to propene

ture separation processes, and the hydrogen could be recycled. Finally, propane could be removed from the propene stream using membranes or other traditional technologies.

A major concern is the source of hydrogen. The basic idea is that the glycerol hydrogenolysis plant may be constructed near a petrochemical plant, where hydrogen is normally available from cracking of naphtha or natural gas reforming. In addition, the propene stream could be straightforwardly directed to the polymerization units, reducing the costs of storage and transportation.

Another possible approach of a biorefinery plant of glycerol hydrogenolysis to propene would involve the use of hydrogen from a renewable source. Presently, the technologies for the production of hydrogen from water or biomass are still not competitive with those based on fossil fuels. Water electrolysis is a well-established technology, but the costs of electricity generation may impair its utilization at high scale. Processes based on the photocatalytic water splitting or photovoltaic decomposition are still low in efficiency, requiring further studies and optimization. Hydrogen can also be produced from biomass gasification, which involves reforming to syngas followed by the shift process. Nevertheless, the process is energy demanding and still cannot compete with the process based on natural gas or coal. Chapter 4 highlighted glycerol reforming to hydrogen, which is a potential biomass feedstock that has been studied for this purpose.

Biomass fermentation is another possibility of producing hydrogen. In fact, even glycerol can be biotechnologically transformed in hydrogen, and this subject has been briefly addressed in Chap. 3. However, the available technologies still require major developments, including the utilization of the carboxylic acids formed as by-products. Thus, a process of glycerol hydrogenolysis completely dependent on renewable sources is somewhat distant from reality, mostly because of the hydrogen source.

6.3.2 *Glycerol Dehydration to Acrolein and Acrylic Acid in the Context of a Biorefinery*

Acrylic acid is used in the production of superabsorbent polymers, paints, adhesives and printer ink, among others. It is industrially produced from the oxidation of acrolein, which in turn is produced through the selective oxidation of propene. Glycerol

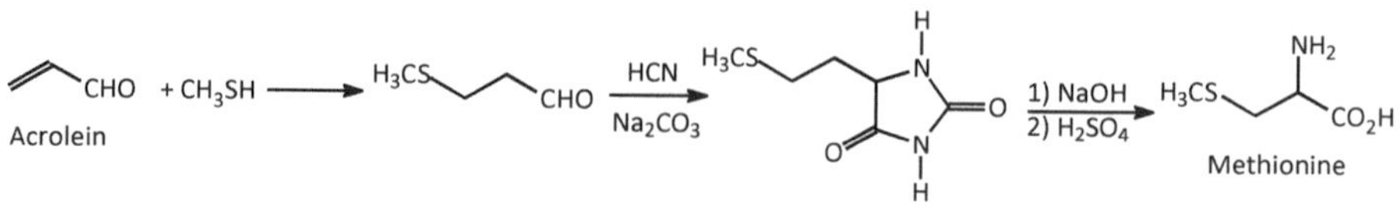

Scheme 6.5 Industrial chemical synthesis of methionine

may be a good renewable raw material for the production of acrolein and acrylic acid, due to its structure with three carbon atoms.

Dehydration of glycerol may afford acrolein as the main product. However, conditions must be carefully chosen to avoid secondary reactions and deactivation of the acidic catalyst. Acetol (hydroxy acetone) may also be formed, upon dehydration of a primary hydroxyl group. This subject has been discussed in Chap. 4. Besides its importance in the fabrication of acrylic acid, acrolein is the raw material for the production of the amino acid methionine, extensively used for animal feed in livestock production [21], through a multistep synthetic route (Scheme 6.5).

Evonik is the main producer of methionine through the chemical synthesis route, with an installed capacity of 580,000 tons per year. A biotechnological pathway has been recently developed [22], and an industrial plant in Malaysia is under construction. A biorefinery aimed to produce methionine from glycerol would involve the acid-catalysed dehydration to acrolein. Nevertheless, the other steps involve hazardous and fossil-derived chemicals and would not be the best choice for the valorisation of the glycerol of biodiesel production in a biorefinery and green chemistry context.

Once an acidic catalyst and reaction conditions to dehydrate glycerol has been developed and optimized, the acrolein formed could be used in the traditional oxidation process to acrylic acid, based on Molybdenum and Vanadium oxide catalysts. Notwithstanding, the costs may be higher compared with the fossil-source pathway. On the other hand, the development of a one-step process, where the glycerol molecule is firstly dehydrated to acrolein, which is subsequently oxidized to acrylic acid would be competitive. This process would be competitive with the traditional route, because of the reduction of the capital and operational expenditures, and requires bifunctional catalysts, with acid and redox properties. This subject has been discussed in Chap. 4. Although several studies on the oxidehydration of glycerol to acrylic acid has been reported in the literature, the selectivity is still low to moderate, pointing out that further research is still required before an industrial plant could be built. A biorefinery based on this technology would be highly interesting, because acrylic acid is a commodity of great value and used in the manufacture of important end-use products.

References

1. Pico MP, Romero A, Rodríguez S, Santos A (2012) Etherification of glycerol by tert-butyl alcohol: kinetic model. Ind Eng Chem Res 51:9500–9509
2. Pinto BP, Lyra JT, Nascimento JAC, Mota CJA (2016) Ethers of glycerol and ethanol as bioadditives for biodiesel. Fuel 68:76–80

3. Wessendorf R (1995) Glycerinderivate als kraftstoffkomponenten. Erdoel Kohle Erdgas Petrochemie 48(3):13
4. Nanda MR, Yuan Z, Qin W, Ghaziaskar HS, Poirier MR, Xu CC (2014) Catalytic conversion of glycerol to oxygenated fuel additive in a continuous flow reactor: process optimization. Fuel 128:113–119
5. Shirani M, Ghaziaskar HS, Xu CC (2014) Optimization of glycerol ketalization to produce solketal as biodiesel additive in a continuous reactor with subcritical acetone using Purolite® PD206 as catalyst. Fuel Process Technol 124:206–211
6. Oliveira PA, de Souza ROMA, Mota CJA (2016) Atmospheric pressure continuous production of solketal from the acid-catalyzed reaction of glycerol with acetone. J Braz Chem Soc 6:1832–1837
7. Mota CJA, da Silva CXA, Rosenbach N Jr, Costa J, da Silva F (2010) Glycerin derivatives as fuel additives: the addition of glycerol/acetone ketal (solketal) in gasolines. Energy Fuel 24:2733–2736
8. García E, Laca M, Pérez E, Garrido A, Peinado J (2008) New class of acetal derived from glycerin as a biodiesel fuel component. Energy Fuel 22:4274–4280
9. Ndaba B, Chiyanzu I, Marx S (2015) N-butanol derived from biochemical and chemical routes: a review. Biotechnol Rep 8:1–9
10. Silva PH, Gonçalves VLC, Mota CJA (2010) Glycerol acetals as anti-freezing additives for biodiesel. Bioresour Technol 101:6225–6229
11. Honda M, Kuno S, Sonehara S, Fujimoto K, Suzuki K, Nakagawa Y, Tomishige K (2011) Tandem carboxylation-hydration reaction system from methanol, CO_2 and benzonitrile to dimethyl carbonate and benzamide catalyzed by CeO_2. ChemCatChem 3:365–370
12. Tomishige K, Kunimori K (2002) Catalytic and direct synthesis of dimethyl carbonate starting from carbon dioxide using CeO_2-ZrO_2 solid solution heterogeneous catalyst: effect of H_2O removal from the reaction system. Appl Catal A 237:103–109
13. Eta V, Mäki-Arvela P, Wärna J, Salmi T, Mikkola JP, Murzin D (2011) Kinetics of dimethyl carbonate synthesis from methanol and carbon dioxide over ZrO_2–MgO catalyst in the presence of butylene oxide as additive. Appl Catal A 404:39–46
14. Marciniak A, Mota CJA (2017) Methyl trihaloacetic esters as efficient and sustainable water suppressors in the synthesis of dimethyl carbonate from CO_2. Chem Select 2:1808–1811
15. Hu ZC, Zheng YG, Shen YC (2011) Use of glycerol for producing 1,3-dihydroxyacetone by *Gluconobacter oxydans* in an airlift bioreactor. Bioresour Technol 102:7177–7182
16. Ma L, Lu W, Xia Z, Wen J (2010) Enhancement of dihydroxyacetone production by a mutant of *Gluconobacter oxydans*. Biochem Eng J 49:61–67
17. Hekmat D, Bauer R, Fricke J (2003) Optimization of the microbial synthesis of dihydroxyacetone from glycerol with *Gluconobacter oxydans*. Bioprocess Biosyst Eng 26:109–116
18. Mota CJA, Gonçalves VLC, Fadigas JC, Gambetta R (2009) Preparation of heterogeneous catalysts used in selective hydrogenation of glycerin to propene, and a process for the selective hydrogenation of glycerin to propene. WO2009155674A1
19. Mota CJA, Gonçalves VLC, Fadigas JC, Gambetta R (2011) Preparation of heterogeneous catalysts used in selective hydrogenation of glycerin to propene, and a process for the selective hydrogenation of glycerin to propene. US20110184216A1
20. Mota CJA, Gonçalves VLC, Mellizo JE, Rocco AM, Fadigas JC, Gambetta R (2016) Green propene through the selective hydrogenolysis of glycerol over supported iron-molybdenum catalyst: the original history. J Mol Catal A 422:158–164
21. Wilkee T (2014) Methionine production – a critical review. Appl Microbiol Biotechnol 98:9893–9914
22. Fremy G, Barre P, Kim S-Y, Son SK, Lee SM (2013) Patent to Arkema France-CJ Cheil Jedang China: preparation process of L-methionine. WO2013/029690(A1)

Index

C.J.A. Mota et al., *Glycerol*, DOI 10.1007/978-3-319-59375-3